Daniel Ammen

The American Inter-Oceanic Ship Canal Question

Daniel Ammen

The American Inter-Oceanic Ship Canal Question

ISBN/EAN: 9783337317959

Printed in Europe, USA, Canada, Australia, Japan

Cover: Foto ©berggeist007 / pixelio.de

More available books at **www.hansebooks.com**

THE

AMERICAN

Inter-Oceanic Ship Canal

QUESTION.

BY

Rear-Admiral DANIEL AMMEN,
U. S. NAVY.

PHILADELPHIA:
L. R. HAMERSLY & CO.,
1510 CHESTNUT STREET.
1880.

Entered according to Act of Congress, in the year 1879,
BY L. R. HAMERSLY & CO.,
In the Office of the Librarian of Congress, at Washington, D. C.

CONTENTS.

INTRODUCTORY.

THE SUFFICIENCY OF OUR INFORMATION IN RELATION TO THE TOPOGRAPHY OF THE AMERICAN ISTHMUSES.

THE FEASIBILITY OF AN INTER-OCEANIC SHIP CANAL VIA LAKE NICARAGUA AS A COMMERCIAL QUESTION.

THE PRESENT ASPECTS OF THE INTER-OCEANIC SHIP CANAL QUESTION.

APPENDIX:

PROCEEDINGS IN THE GENERAL SESSION OF THE CONGRESS IN PARIS, MAY 23, AND IN THE TECHNICAL COMMISSION, MAY 26, 1879.

REPORTS OF REAR-ADMIRAL DANIEL AMMEN, U. S. NAVY, TO THE SECRETARY OF STATE, JUNE 21, 1879.

REPORT OF CIVIL ENGINEER A. G. MENOCAL, U. S. N., TO THE SECRETARY OF STATE, JUNE 21, 1879.

INTRODUCTORY.

Circumstances which occurred nearly a quarter of a century ago, briefly mentioned in the following paper read before the American Geographical Society of New York, fixed my attention on the question of the possibility of the construction of an inter-oceanic ship canal across this continent.

Owing to the approaching struggle into which we soon passed, and the perturbation which preceded and followed it, no possibility of making further explorations occurred for years.

In the early part of the winter of 1866 I was in Washington, in command of a vessel of war. At my request, Rear-Admiral Chas. H. Davis, Superintendent of the Naval Observatory, prepared a map on a large scale of the narrow part of this continent for General Grant, with whom I had the pleasure of discussing what was then known and what was still in doubt respecting the topography of that region. This led to one or more visits to Mr. Seward, then Secretary of State, whose reception of the subject of making further explorations at that time made General Grant averse to seeing him further in relation to this matter.

Mr. Conness, then Senator from California, offered a Resolution calling for information from the Superintendent of the Naval Observatory, which was given during that year, 1866. He continued his interest in this subject, and finally obtained an appropriation for making inter-oceanic surveys in the winter of 1869. In the mean time General Grant had been elected President, and did not fail to do all in his power, through his subordinates, in forwarding surveys, which, when terminated, left no part of the topography of the Isthmus in doubt, so far, at least, as their possibility for the construction of a ship canal was concerned.

Under a Resolution of Congress, as President, he appointed a Commission on March 13th, 1872, consisting of the Chief of Bureau of Engineers, U. S. A., the Superintendent of the Coast Survey, and the Chief of Bureau of Navigation, U. S. N., to examine into, make suggestions, and report upon the subject.

This Commission found further information indispensable; it informed the President that a close instrumental examination of the Isthmus of Panama in the immediate region of the Panama railroad was necessary to a full consideration of the subject. The President directed an immediate execution of this work, which was completed with the least possible delay. The Commission then decided, and reported to the President on the 7th of February, 1876, in the following terms:—

"That the route known as the 'Nicaragua route' possesses, both for the construction and maintenance of a canal, greater advantages, and offers fewer difficulties from engineering, commercial, and economic points of view, than any one of the other routes shown to be practicable by surveys sufficiently in detail to enable a judgment to be formed of their relative merits as will be briefly presented in the appended memorandum."

Owing to reasons not fully known to me, and difficulties which I will not surmise, no actual progress was made in bringing about or furthering the construction of the work previous to the expiration of the Presidential term in March, 1877, when General Grant went out of office.

To my personal knowledge, General Grant did not at that time, nor until recently, have any disposition to participate in the work of construction of a ship canal. When great difficulties arose, in propositions for making a canal at the ocean-level at Panama, apparently without regard to any commercial consideration of the question, or of the permanency of the work, he probably regarded it of such great importance to our future commerce, and that of the world generally, as to express a willingness to aid actively in its construction on a route which presents relatively economical conditions for construction, and permanency, as far as it is possible in such works.

Whether this canal will or will not be made depends upon the appreciation of its merits; not upon what is *possible*, but what is certain of realization with a given amount of expenditure. To make its construction an assured fact, its appreciation must be by the moneyed interests of the world, and notably by those of the United States and of Great Britain, whose interests in the construction of such a work are so largely preponderant to those of other peoples and nations.

From page 608 of the "Proceedings of the Royal Geographical Society of London" for September, in noting the proceedings of the Geographical Society of Paris, I quote the following:—

"M. de Lesseps announces that the subscriptions for 800,000 shares, at 500 francs each, of the Panama Inter-Oceanic Canal Company, would be opened on the 6th of August. He expressed himself astonished, and even disappointed, that the project was advancing so smoothly. A little serious opposition would have been agreeable to him."

Whilst not sharing his sentiments as to the agreeability of contentions, I should say that if an opposition to him, or rather his project, from a preference of another route for a ship canal, is agreeable to him, so much the better. We may well wish him all the success that the merit of his project possesses on physical conditions only, and may regret any injury to it from "diplomatic or political intervention," should it be supposed desirable or possible.

SURVEYS AND RECONNOISSANCES FROM 1870 TO 1875 FOR A SHIP CANAL ACROSS THE AMERICAN ISTHMUS.

BY COMMODORE DANIEL AMMEN.

WASHINGTON, D. C., October 21, 1876.

HON. CHAS. P. DALY, *President Am. Geographical Society, N. Y.*

DEAR SIR:—In reply to the request of the Society for information in regard to the recent surveys which have been executed by the Government of the United States for a Transcontinental Ship Canal across the American Isthmus, I very cheerfully send you the enclosed communication, placing it at the disposal of the Society. In presenting the paper relating to these surveys, a brief explanation may, however, be expected from me by at least some of your members.

More than twenty years ago I was attracted to the consideration of this subject by the then published accounts of the coincident attempts by the English, the French, and ourselves, in the vicinity of Caledonia Bay, to discover the pretended Cullen route,—a route disproved by those three parties, but which, having been again insisted upon as feasible, was recently shown by the American Expedition (1870) to be a physical impossibility, by reason of the elevation of the watershed adjacent to and across the "divide" near Caledonia Bay, the chief streams of which water-shed flow into the Pacific.

At the date referred to, the English, working from the Savannah River, ran a line of levels towards Caledonia Bay and reached the waters of the Sucubti, the stream north and west of the Caledonia Bay range of mountains, at a height sufficient to show the impracticability of the route.

The French seem to have abandoned their work without producing instrumental results indicating in any degree a hope of success.

The American party, under Lieut. Strain of the Navy, ascended the mountain range from the bay, reached the Sucubti on the Pacific slope, and without the use of instruments of precision, followed the tortuous stream to the Chucunaque, and made their way down that still more tortuous stream, with the loss of more than half of their number by starvation. This was the natural result of not providing proper outfit, and carefully husbanding their provisions. No positive knowledge was gained by our expedition, except the necessity that future explorers should be judiciously provisioned and equipped with the means of securing the best instrumental results.

After two days' descent of the Sucubti, Strain's party, when encamped upon an island, had supposed they had heard the evening gun of the "Cyane," the vessel which they had left anchored in Caledonia Bay. This supposed fact, seemingly without significance to them, attracted my attention; for if it were a fact, it would appear to indicate the existence of a low line of levels between that point and the waters of Caledonia Bay. It seemed to me not at all likely that the sound was deflected up a mountain side and again descended through the valley beyond.

I was led to consider more fully the probability of a low line of levels near the point referred to, and further to study the question of meeting the formidable obstacles besetting explorers in this almost impassable region, and of securing sufficiently positive, conclusive knowledge of the country, to establish, in relation to all the watersheds, the practicability of a transcontinental ship canal—or the reverse.

Presenting my views, in 1856, to Mr. Toucey, then Secretary of the Navy, but failing to receive his countenance and support, I went to the Pacific Ocean on board of one of our vessels of war, and did not return until 1860, when I wrote out briefly my project for exploring the entire region necessary to be examined, in a paper which this Society did me the honor to read on the 7th of June of that year. This project has, in fact, formed the basis of our surveys and explorations, modified as has been found necessary by the intelligent and able officers who have actually executed the work.

At the time when the paper was read to your Society the political condition of our country was disturbed, and the years of civil war which followed indefinitely postponed even the consideration of this most important subject. On my return from the Asiatic station in April, 1869, I was gratified at finding that appropriations had been made for transcontinental ship canal surveys, and that General Grant, then President, was initiating a comprehensive examination and sufficient surveys of the extensive region involved. Their full execution has required years of labor and the employment of large, well-equipped parties, as will hereafter be shown.

For the past five years, during which I have been Chief of the

Bureau of Navigation, the Secretary of the Navy has honored me by directing the Bureau to give special attention to the selection of most efficient officers for this work; to look closely to the proper supply of articles of subsistence, and for the best instruments found by experience to be suitable; to formulate orders for his examination and approval; to examine closely the results of surveys; and to supply whatever deficiencies might be found to exist for the full investigation and determination of this question.

Since the appointment by the President of the Commission to investigate and report upon a transcontinental ship canal route,* all orders and instructions for surveys in progress have been, in effect, in accordance with the wishes and requirements of that Commission; at their instance a close instrumental survey and actual location of a route was made on the Isthmus of Panama, and a further examination of the Chepo-San Blas route from the Pacific coast.

The surveys were at length completed, as will be referred to in detail. They were satisfactory to the Commission, and I can assure you that their execution was no holiday work. In every case where tentative lines only were prosecuted, this was done instrumentally to a point developing impracticability, or else a manifest inferiority for construction as compared with other lines found more favorable. On the latter, instrumental locations for a canal were made, and plans and approximate estimates of construction prepared.

I cite, in this connection, a few paragraphs from my Report to the Secretary of the Navy, of October 26, 1875 [Annual Report of Secretary of the Navy for the year 1875, pages 60, 61]:—

"The arduous work which has been carefully prosecuted for five "seasons by two or more parties, from the Isthmus of Tehuantepec to "twenty or more miles south of the mouth of the Napipi, on the "River Atrato, is at length satisfactorily accomplished.

"It is the duty of this Bureau to acknowledge the ability and energy "of the different officers who have been in command, and the untiring "zeal and faithful and intelligent exertions of their subordinates. "The precautions of those in command are shown in the fact that not "one officer or man has succumbed to climatic influences, though "many doubtless carried the seeds of disease and earlier death away "from their field of operations. No case of bad conduct in either

---

* The Commission, appointed March 13, 1872, was ordered to consist of the Chief of Engineers, U. S. A., the Superintendent of the Coast Survey, and the Chief of the Bureau of Navigation.

"officer or man engaged on this work has came to the knowledge of
"the Bureau." * * *

" By tentative surveys, following in the main up the various valleys
" on both coasts, until reaching heights and distances apart that would
" make the different water-sheds between the points named inferior to
" other points already known, the process of elimination was com-
" pleted. It was a long, laborious process, taxing the endurance of the
" officers and men.

" Since my last Report, at the request of the Commission appointed
" to consider and report upon the inter-oceanic canal, by your order, a
" careful survey of the Isthmus of Panama was made, the computa-
" tions completed, and the whole placed before the Commission.

" A reconnoissance on the west coast was also made of the Rio
" Chepo and the San Blas route, where the tide-water of the two
" oceans approach more nearly than at any other point. This work
" was executed by Commander E. P. Lull, U. S. N., and junior naval
" officers, aided by Civil Engineer A. G. Menocal, U. S. N.

" At the instance of Commander T. O. Selfridge, who had executed
" the former work on what is known at the Napipi route, the Depart-
" ment directed the fitting out of another expedition to make an actual
" location of an inter-oceanic canal along this line.

" This work was assigned to Lieut. F. Collins and junior naval
" officers.

" The work has been successful accomplished, the computations
" made, and placed before the Commission.

" So careful and minute has been the examination of the different
" water-sheds up to the point of manifest inferiority to other known
" points, that no doubt now exists as to the approximate labor neces-
" sary in the construction of an inter-oceanic ship canal at several
" points. It is proper to add, that the most careful and elaborate
" surveys would necessarily have to be made in advance at any point
" heretofore examined before commencing the construction of an inter-
" oceanic ship canal, and that these surveys could only ameliorate the
" labor and cost of construction, inasmuch as *the locations, as given,*
" *are actual throughout their length,* and would only be changed when
" an advantage would be gained by doing so."

In view of these conclusions, which I hope to establish fully with those who will re-examine the various surveys that will be hereafter summarized, I have read with some surprise the postulates recently and widely published by M. Leon Drouillet, engineer, and member of a " Commission of Commercial Geography of Paris," lately formed under the sanction of the French Société de Géographie—postulates on this subject thus seemingly endorsed by that learned and distinguished body.

Through the kindness of M. Drouillet, I have been favored with a copy of the proceedings of the " French Section of the International Committee for the Exploration of the American Isthmus," and also with a pamphlet, of which this gentleman is the author, elaborating a plan for such an international exploration.*

In the pamphlet referred to, M. Drouillet, when urging the necessity of an International Exploration of the American Isthmus, sets out with the following postulates:—

" Le problème de la Navigation inter-Océanique est actuellement " insoluble par suite de l'insuffisance des données géographiques et des " contradictions flagrantes qui existent dans ces données; insuffisance " et contradictions qui ne permettent point à l'ingénieur l'étude appro- " fondie d'un project définitif."

[" The problem of inter-oceanic navigation is, at present, incapable " of solution on account of the insufficiency of geographical data, and " of the flagrant contradictions which exist in these data,—an insuffi- " ciency and contradictions which do not permit the engineer to study " profoundly a definite project."]

On the strength of these assertions, with the seeming approval of the Geographical Society of Paris, an appeal is made to the learned societies of the world, and to all the powers interested, to lend their aid to a " general and serious exploration of the Isthmus."

In view, therefore, of the long series of elaborate explorations and reconnoissances lately made by the United States, reported upon by the Commission to the President, and *accepted as satisfactory by him*, it seems proper to present what has been really done by us, and to leave to the good judgment of those societies and interested powers whatever action seems to them necessary or advisable.

The demand for a re-survey is rested upon two principal grounds:—

1st. That the data at hand are insufficient.

---

* " Société de Géographie et Commission de Géographie Commerciale de Paris, Sec-
" tion Française du Comité International d'etude pour l'éxploration de l'Isthme Américain
" en vue du percement d'un Canal inter-Océanique.
" Procès verbal de la séance du 11 Mai, 1876."
" Les Ishthmes Américains—Projet d'une exploration Géographique Internationale des
" terrains qui semblent présenter le plus de facilités pour le Percement d'un Canal Mari-
" time inter-Océanique. Par M. Léon Drouillet, Ingénieur Membre de la Société de
" Géographie et de la Commission de Géographie Commerciale de Paris."

2d. That what is available is flagrantly contradictory.

Let us consider these two assertions separately.

As for the sufficiency of the data at hand, without, at present, going beyond the work executed for the most part during the past six years by the United States alone, we may point to the following not inconsiderable sources of reliable information respecting every part of the isthmus, of any promise, for a canal—from Tehuantepec to the Napipi River, in South America.

Of our surveys and reconnoissances the following is a list in the geographical order from the north and west to the south and east, in regard to which list it is to be specially noted that every survey and reconnoissance was made with instruments of precision, unless mention to the contrary is herein made.

All lines upon which calculations have been founded were run by compass and chain, or transit and chain, or by gradienter and stadia-rod, the barometer being relied upon only to fill in the topography on either side of the main line.

1. Instrumental reconnoissance of the Isthmus of *Tehuantepec*, by Capt. R. W. Shufeldt, U. S. N., 1872.

2. Examination, survey, and definite instrumental location of an inter-oceanic canal route from the vincinity of *Greytown via Lake Nicaragua*, and thence via the Rios del Medio and Grande to Brito, by Commander E. P. Lull, U. S. N., 1872 and 1873; with some preliminary operations by Commander Chester Hatfield, U. S. N., in 1872.

3. Examination, survey, and definite instrumental locations of an inter-oceanic canal route from *Navy Bay to Panama*, by Commander E. P. Lull, U. S. N., 1875.

4. Examination and surveys from the *Gulf of San Blas* towards the River *Chepo*, by Commander T. O. Selfridge, U. S. N., 1870; and

Reconnoissance from the waters of the *Chepo* toward the *Gulf of San Blas*, by Commander E. P. Lull, U. S. N., 1875.

5. Several tentative instrumental lines in the vicinity of *Caledonia Bay*, across the Cordilleras to the waters of Sucubti and Morti rivers, tributaries to the *Chucunaque*, by parties under the direction of Commander T. O. Selfridge, U. S. N., 1871.

6. A barometrical reconnoissance of the so-called "*De Puydt Route*," by way of the Tanela River between the Tuyra and the Atrato, by a party under the direction of Commander T. O. Selfridge, U. S. N., 1871.

7. Tentative instrumental lines by the so-called "*Gogorza Route*,"

from the eastern coast *via* the *Atrato, Cacarica,* and Peranchita rivers, and from the west coasts *via* the Tuyra and Cué rivers across the "divide," by parties under the direction of Commander T. O. Selfridge, U. S. N., 1871.

8. An instrumental examination of what is known as the "*Truando Route,*" by Lieuts. Michler, U. S. A., and Craven, U. S. N., 1856–57.

An instrumental reconnoissance of the *Napipi* and *Cuia* rivers, including a reconnoissance of the Atrato River to the town of Quibdô, by parties under the direction of Commander T. O. Selfridge, U. S. N., 1871 and 1873.

10. Tentative examinations and definite instrumental location of an inter-oceanic canal route by way of the *Napipi* and *Doguado* rivers, by Lieut. Frederick Collins, U. S. N., 1875.

The results of these several explorations will now be briefly noted in the same order:—

1. Tehuantepec.—*Indisputably inferior to other known points.* Number of locks required, 140. Length of canalization, 144 miles.

2. *Nicaragua.*—A summit of 107.6 feet; length of canal requiring excavation, 61.75 English miles; slack-water navigation by means of dams in the bed of the San Juan River, from the mouth of the San Carlos to Lake Nicaragua, a distance of 63 miles. Lake navigation for 56.5 miles to Virgin Bay; and thence *via* the valleys of the Rio del Medio and Rio Grande to Brito.

This plan involves the construction of four dams having an average height of 29.5 feet, and an aggregate length of one thousand three hundred and twenty (1320) yards; and of twenty locks of an average lift of ten and twenty-eight hundredths (10.28) feet each. It also involves the construction of two harbors of sufficient extent to insure, at least, a smooth and safe entrance into and exit from the canal.

It is worth remarking, that M. Drouillet, in presenting the fifteen projects in this vicinity, does not distinctly describe this route (projected after a careful instrumental survey, involving several tentative lines from Lake Nicaragua to the Pacific); nor would a reader of the pamphlet referred to assign this line as above presented to any one of the fifteen projects given in it. This leads to the supposition that he has given the preference to some of the less exact surveys or supposititious pretensions quoted as examined; and this belief is entirely verified by the fact that he gives the actual height of Lake Nicaragua above the sea level as 37 metres, which is thirteen and six-tenths (13.6) English feet in excess of the true elevation (as presented by our careful instrumental surveys)—in excess even of the elevation to which the mean elevation of the surface of the lake is to be raised and maintained by a dam.

3. *Panama.*—This survey, executed, as has been said, at the request of the Commission appointed by the President to investigate the whole question of a ship canal, made an actual location along an entire route. Maps, plans, and estimates for excavation and construction have been carefully prepared, as upon the Nicaragua route, and on a common basis of cost for like labor. The Report of the survey published in the Appendix to the Report of the Secretary of the Navy for 1875 does not appear to have been in M. Drouillet's possession.

4. *San Blas.*—The surveys of Commander Selfridge from the east coast, and those of Commander Lull from the western, demonstrated that there is no practicable route between the Gulf of San Blas and the waters of the Chepo, even with a tunnel of eight (8) miles, although between these points the tides approach each other from the two oceans more nearly than elsewhere.

5. *Caledonia Bay.*—The tentative instrumental lines from the northern and southern parts of Caledonia Bay across the "*divide*" to the elevated beds of the Morti and Sucubti rivers showed, for the second time, that the information of Edward Cullen was an invention.

The line from the southern extremity of Caledonia Bay crossed the "*divide*" at an elevation of twelve hundred and fifty-nine (1259) feet, and struck the bed of the Sucubti at a height of five hundred and fifty-three (553) feet, thus precluding the possibility of any pass under that altitude above the point reached on the Sucubti.

The line from the northern extremity of the bay up the valley of the Sassardi and across the "*divide*" to the Morti crossed at an altitude of eleven hundred and forty-eight (1148) feet, and no indication of any pass under one thousand (1000) feet could be discovered.

This line is marked by M. Drouillet for re-examination.

6. *De Puydt's Route.*—The exact line advocated by De Puydt, as obtained from a gentleman who had accompanied him, was followed for some thirty-three (33) miles. At this distance an elevation of six hundred and thirty-eight (638) feet had been reached, while the mountains of the divide were plainly visible beyond. Three mercurial mountain barometers were used; one at the sea-level was observed at short intervals during the whole reconnoissance, the other two were carried by the party; bench-marks were established at convenient distances, one barometer remaining at each bench until another had

reached the next, and until sets of differential observations had been obtained.

7. *The Atrato-Tuyra Route.*—The tentative instrumental lines from the east and the west coasts, which were run in the examination of this supposed route, established the fact that Hellert, La Charme, and Gogorza were pretenders—were it indeed necessary to establish this in the case of those who have done no more than make unsupported assertions. Our regular line of survey—by way of the Atrato and Peranchita rivers on the east, and the Tuyra and Cué rivers on the west—crossed the "divide" at an altitude of 712 feet; while a little further north Capt. Selfridge crossed at a height of 400 feet, as estimated from rough observations with his pocket aneroid.

M. de Gogorza claims that Capt. Selfridge's examinations did not cover his proposed route; but it will always be possible for him and other authors of brilliant but vague projects to make this complaint regarding any expedition not led by themselves. Whether the exact route proposed by M. Gogorza was followed in this case or not, it is certain that the explorations were sufficiently extensive to show that the whole country, on the Pacific side of the divide especially, is a net-work of high hills, which feature, taken in connection with the extensive swamps on the Atlantic side, is sufficient to condemn the route, independently of the height of the dividing ridge.*

(This locality, with two preceding ones, involving also the region of a third, comprise points specially noted by M. Drouillet for examination; he thus entirely ignores the joint attempt by the English, the French, and ourselves on the latter route in 1854, and the recent instrumental disprovement of it, with the others, *by us.*)

---

* Since writing this paper, the pamphlet and map very recently published in Paris by M. de Gogorza have come into my possession. In this pamphlet—" Canal Interoceanique sans écluses ni Tunnels" (!)—M. Gogorza asserts that Commander Selfridge's surveys support his own, as far as they were made over the same ground. This is an ERROR. Commander Selfridge gives the height of the mouth of the river Paya at one hundred and forty-four (144) feet, and the height of Paya village at two hundred and fifty (250) feet. M. Gogorza gives the same height for the mouth of the Paya, but is silent as to the heights in ascending to the village of Paya, 20 miles above, following the sinuosities of the stream, and does *not give the height of that village at all.* He contents himself with asserting that, at a distance of miles beyond the village, at the summit-level, the height is only fifty-eight metres (58),—one hundred and ninety feet,—that is to say, *sixty feet below the village!*

The altitude of the mouth of the Paya River itself, as given by himself, and on better authority, contradicts flatly his assertion that a ship canal, without locks or tunnels, can be located betwen the summit-level, the village of Paya, and the mouth of the Paya River. He terminates his canal at the Isla de Lagartos, but does not locate that significant island.

9 and 10. *The Atrato-Napipi Route.*—This was examined first by parties under the direction of Commander Selfridge, and afterwards by Lieut. Collins. By the last-named officer a definite instrumental location for a canal was made; the question of additional water supply from the Cuia was investigated, and calculations for excavation and construction framed on a common basis for like labor as for Nicaragua and Panama. The report of this survey, without maps and plans, is to be found in the Appendix to the Report of the Secretary for 1875. The lack of appropriation for publishing this report, and that made by Commander Lull on Panama, *in full*, is regretted.

These surveys are not named by M. Drouillet in his list of authorities.

These repeated and laborious surveys certainly indicate the continued interest which the United States has taken in the construction of a canal. This interest dates back, indeed, to the administration of Mr. Jefferson, and its appreciation by the Congress of the United States was shown as early as 1835, by an elaborate report in the House of Representatives; as subsequently by various official inquiries and American treaties. (See Report No. 145, Ho. Rep. 30th Congress, 2d Session, *et al.*)

If necessary, not a few other American authorities might be cited, such as those of Trautwine, Kennish, Porter, Totten, and Childs, employed by private American enterprise, as affording reliable information within the limits claimed; but it would appear that the sufficiency of the data is already manifest, provided their authenticity is unquestioned: and this brings me to the second postulate of M. Drouillet, that " the data at hand are flagrantly contradictory."

But here I repeat that our surveys have been pursued for several years by officers of well-established reputation and ability, and by full and competent scientific staffs, with every advantage of outfit, of instruments and stores; and, in the latter surveys, with the additional advantage of the experience possessed by the principal officers,—an experience to be acquired only in the field.

The scientific staff of the first expedition of Commander Selfridge numbered thirty-five (35) members, including astronomers, geologists, mineralogists, topographical and hydrographical engineers, telegraphers, photographers, and others. The men attached to this expedition, exclusive of natives employed as laborers, numbered about 300. Three ships-of-war were also attached to the survey,—two on the Atlantic side and one on the Pacific side.

The scientific staff of the second expedition of Commander Selfridge numbered thirty (30) members, exclusive of the officers of the U. S. S. "Nipsic" and "Resaca," both of which, with the U. S. S. "Guard," were attached to the expedition.

The Tehuantepec and Nicaragua expeditions were equipped with like liberality.

Able officers of the U. S. Coast Survey and Civil Engineers were associated with the commanding officers in these various expeditions; notably Messrs. Sullivan, Mosman, Ogden, Merinden, and Blake, all of them distinguished Coast Survey officers, trained in the severely correct methods of that service, were with commander Selfridge; A. G. Menocal, Civil Engineer, U. S. N., with Commander Lull; A. E. Fuertes, Civil Engineer, with Capt. Shufeldt. Commander Lull and Lieut. Collins had served with Commander Selfridge in the earlier expeditions in Darien, and many of the officers subsequently associated with them had also seen service in the same way.

The work assigned to the expeditions thus equipped was laid out by careful and ample instructions from the Navy Department, and was satisfactorily performed; the results obtained are believed to be all that the nature of the conditions rendered possible.

None of these extended surveys conflict in any degree with each other or with other partial surveys or reconnoisances which have been at times undertaken by private American enterprise. If any authentic instrumental surveys or proper tentative lines in the possession of M. Drouillet disprove or contradict any one of our surveys, this would certainly be of profound interest to the learned societies of the world, and afford for them foundation for further projects of exploration, however little they are considered necessary by those who have gone through these repeated practical labors and experiences in the gloomy fastnessness of the great American isthmus. Until, however, such *authentic* contradictory data can be shown, it must appear that the "flagrant contradictions" asserted to exist arise from a want of placing merited confidence in the surveys of the United States. If the unsupported statements of men who discover the proper site for an inter-oceanic canal by "observing the flight of low-flying Pisisi ducks," or who obtain their altitudes "by the velocity of mountain-streams," or the boiling-point of water merely, or who are confident of a continuous depression from the mere aspect of the forests, as seen from on board ship, or from having observed an "inclination of the ground to be scarcely perceptible,"—if these deceptive appearances, so well recognized by travelers, some of which were strongly noted in this very connection by Humboldt when describing his ascent from

Callao to Lima, are to be placed alongside of official Government surveys, then certainly "the flagrant contradictions" must be expected, and will certainly exist, if even the new, general, and "*serious*" survey now claimed to be necessary is undertaken and completed.

The deceptive appearance of the mountain-ranges from the sea, which has misled so many, was thus noted by Lieut. Michler, U. S. A., in his Report of 1856–57: "In looking back from the ocean upon the country through which the travelers had recently passed, the depression in the Cordilleras becomes plainly visible. It seems to lose its mountainous character entirely. * * * One can easily, therefore, conceive why a preference should have been shown to this section by those interested in the construction of a canal." And the common experience of our officers on the isthmus has been, that wherever a line of low elevation has been affirmed to exist on the strength of the authority of "old Spanish maps or documents," or on the information of "intelligent persons residing in the vicinity," or "through conversations with the natives," *there* an elevated, forbidding range of mountains or hills has been found.

Our surveys have been undertaken and conducted with a view to ascertain the relative practicability of all possible canal routes. It is not affirmed that they are sufficiently extensive and minute at all points to afford the engineer full data for locating a canal, and for estimating its approximate cost. Actual instrumental locations of determinate lines throughout were made at three points only,—at Nicaragua, Panama, and the Napipi. The tentative lines in other places were carried only sufficiently far to demonstrate their impracticability or manifest relative inferiority; thus eliminating, however, all such territory from the canal problem.

If it is in the plan of M. Drouillet, or of others, to procure the precise data called for by the engineer on each of the pretended, or of the real lines of promise for a canal, there will certainly be need by such parties of the most extensive co-operation in every particular, which is invited in the publication referred to.

The natural conditions of the American isthmus will be found widely different from those of Suez, to which constant reference is made. One is a region of extraordinary rainfall—the other of extreme dryness; the one covered with impenetrable and interminable forests —the other wholly denuded; the one a region of steep escarpments and water-sheds, where every ravine, many times during the year, becomes a river of rapid waters rushing wildly to the sea and bearing huge masses of silt, giant bowlders, and fallen trees—the other simply a sandy level plain. If the existence of any narrow American valley, many miles in length, between the seas, *be* admitted, and a canal without locks be supposed to be located therein, *it must become the ultimate drainage of that whole tropical valley.* By what human power could it be kept clear of the debris swept into it by every heavy rainfall along its entire length?

Let such low valleys, however, continue to be pointed out, "for a canal without lock or tunnel," as by M. De Puydt, M. Gogorza, or by whomsoever can hold forth the most brilliant promise; let further search be made by whomsoever feels interested, hopeful, and credulous; and let the work go on, aided by such forces, Governmental or otherwise, as may be furnished: the United States and its learned societies may properly decline co-operation. The question whether the authorities I have quoted are sufficient to determine the location of a transcontinental ship canal is an open one; those who think the authority insufficient may well proceed with whatever surveys they may deem necessary.

In submitting to learned societies what has been done, no indulgence for nationality is desired; nor, on the other hand, can there be a tenable assumption that we are incapable of obtaining results which can be obtained by others, or that we have not the integrity to present them fairly. Surveys tell their own story, and discredit themselves if they are to be discredited. I feel sure that this learned body would not willingly discourage others in the prosecution of further surveys, however unnecessary the Society may consider them, and however unwilling, therefore, itself to participate in them.

In Paris, in August, 1875, it was urged by persons who may, perhaps, be properly styled adventurers, attending the International Geographical Congress, that the Government of the United States had really shown no interest in the subject of a transcontinental American canal, and that our information and surveys amounted to but little. The misapprehension on the first of these points, if it exists, is too apparent to require contradiction; the assumption of the second seems the result of not having examined what the United States have done, or it is the affectation of a belief that we cannot do the work as well as any other people.

So far from the United States being indifferent to the construction of an inter-oceanic canal, for more than fifty years,* as has been

---

* See Correspondence between Mr. Clay and our Chargé, Mr. Williams, and Mr. Canaza, "Minister of the Centre," in April, 1825.

shown, we have endeavored to establish the practicability of the work at the most favorable point; and I venture nothing in asserting that our Government will be anxious to do whatever is proper to aid in the construction of the work on the broadest principles of common benefit to all nations and peoples.

It is asserted that Great Britain would oppose it, under the supposition of its injury to the Suez Canal, in which she has now a pecuniary interest. But when by reference to the map it is seen how readily an American ship canal will bring her into communication with the eastern coast of Australia and New Zealand, relieving her outward-bound voyages of head winds, it is safe to assert that such advantage, with others, would exceed in value yearly her whole interest in the Suez Canal.

The American transcontinental canal will bring Great Britain within easy commercial relations with the entire west coast of America, exchanging the stormy passage around Cape Horn, with head winds, for a short American route with fair winds and good weather; it will make the products of British Columbia and of Central America (where British trade is even now so important) doubly valuable; and it will bring the wheat products of California more fully and competitively into British markets, thus cheapening and making less fluctuating the price of breadstuffs,—an incalculable advantage for the masses of a manufacturing and commercial country. And while this is true, it is equally demonstrable by an inspection of the world-map that the great lines of commercial intercourse and of civilization are distinct, and therefore uninviting any prejudicial rivalry as regards the two—the Eastern and the Western isthmus—ship canals. The Suez Canal is the opened gate for the inland sea route of Europe and North Africa with Southern Asia and its Archipelago: the American isthmus canal invites Europe with our own commerce to the whole west coast of the Americas, to Northern China and Japan, and southwardly to the Australian Continent. Nor can commerce longer forget that not only the drainage of the rivers emptying into the American Mediterranean is of an area greater than that of all the rivers emptying from Europe into the Atlantic, and of all those emptying into the Mediterranean and into the Indian Ocean, but that the valleys of these American rivers are those of different productive zones. The back country essential to commerce exists here, therefore (as Maury showed twenty-five years ago), around the Mexican Gulf and the Caribbean Sea, larger than that around any other sea.

It is said that the overland railroad interests will oppose the construction of a transcontinental ship canal. This will certainly not be the case if they study their own advantage. This year the wheat crop for export at California alone is stated to be in excess of twenty millions (20,000,000) of sacks of 100 lbs. each, *none of which can be sent to the Eastern coast by railroad without a commercial loss.*

The undoubted advantage to a railroad is to favor the most economic means of transport of this great product, and of other gross and valuable products not transported by rail. For by thus promoting their increase (needed for the supply of Europe and of our own Eastern coast) the railroad must surely gain a recompense through the travel consequent upon an increased and healthy population on the Pacific coast, and the transportation of the light and valuable freights that would of necessity then exist to meet their wants. It requires but small comprehension of the situation to appreciate the fact that the construction of a transcontinental canal would build up instead of injuring railroad interests between the two oceans.

By our geographical position and relative proximity to the isthmus and the countries beyond, we have a greater interest than any European power in the construction of this great work. The commerce of the whole world, however, has large interest in it, and, therefore, the cost of its construction, and its profits, as well as other consequent advantages, seem common to all. This necessarily involves a broad neutrality for the canal and its approaches,—a neutrality to be supported either passively or actively, as the nations may best exert their forces.

The correction of misconceptions on this whole subject, and the apparent advantage of stating, at this time, definitely its true condition, has been the object of this paper. In common with many others, I have looked for many years, with much interest, to the development of this problem in a commercial view, which, in fact, involves its realization. No doubt exists now of this commercial practicability. I may add, as a personal conviction, that, however long and seriously the search may be continued for "results" by surveys, nothing can be or will be developed so advantageous as that which the surveys of our Government present for your consideration.

I am very respectfully yours,
DANIEL AMMEN,
Commodore U. S. Navy and Chief of Bureau of Navigation.

INTER-OCEANIC SHIP CANAL ACROSS THE AMERICAN ISTHMUS.

BY REAR-ADMIRAL DANIEL AMMEN, U. S. NAVY.

THE PROPOSED INTER-OCEANIC SHIP CANAL BETWEEN GREYTOWN AND BRITO, VIA LAKE NICARAGUA; ITS FEASIBILITY AS A COMMERCIAL QUESTION, AND ITS ADVANTAGES AS COMPARED WITH OTHER PROPOSED LINES.

[Read before the American Geographical Society, Nov. 12, 1878.]

Two years ago I had the honor to have read before this Society, by its Secretary, a paper upon the sufficiency of the information relating to the topography of this continent, for the consideration and discussion of a transcontinental ship canal. The object was to set forth what was really known concerning the topography of the whole region which might be regarded as affording possible conditions for its construction.

Notwithstanding the laborious and creditable surveys made public before that time, and quoted to show the sufficiency of our information, it was believed by some learned men in Europe that the topography of certain portions of the territory had not been sufficiently developed. Under this belief, with praiseworthy action, they had set on foot a surveying party, well-equipped, under the command of Lieutenant Wyse, of the French Navy, who was *en route* to make projected surveys over the region referred to, at the time of the reading of my paper.

Two seasons have been employed in the execution of these surveys, which have, without doubt, been ably made, and are reliable within the limits claimed by experts in such works; they are sufficient to form the basis of an instrumental location of the proposed work, and to permit an approximate calculation of the labor necessary where construction is within the possibilities of the engineer. These surveys are, in this respect, useful, and especially for the consideration of those who thought the information presented two years ago was insufficient.

It is a fair presumption that the route over which an inter-oceanic ship canal should be constructed can only be named when it is shown that no better one exists. At this time the information is supposed to be all that can be desired by any one, relative to a fair consideration of the subject.

After an examination of the Reports of Lieutenant Wyse, of both seasons, I have the belief that the closing paragraph of my last paper is verified; namely, "That no possible route exists comparable with what had been presented in the surveys made by order of our Government."

The able Reports of Commander E. P. Lull and Civil Engineer A. G. Menocal, U. S. Navy, on the Nicaragua route, are sufficiently full for examination and criticism by the civil engineer or the expert. There has been given, throughout, a careful consideration to that most important, indeed it may be called vital, question in the construction of an inter-oceanic ship canal in that region, an ample and studied provision to prevent any considerable quantity of surface drainage entering the canal, and the feasibility of accomplishing this object on the located route, as compared with other routes, is, in my belief, a most important point in its favor.

The most important physical feature is the existence of Lake Nicaragua, which, when full, is 107.6 feet above the ocean-levels. This is designed to be the summit-level of the canal. The lake has a superfice of 2800 miles, with a surface drainage falling into it roughly estimated at 9450 miles, which furnish an outflow of water by the San Juan River twenty times as great as the commerce of the world could require in the lockage of vessels passing through the canal.

Even this superabundant water supply without the cost of a feeder is not the most beneficent feature; it is in the equalization of floods from so large a superfice, so that in reality the surface level of the lake may be said to change almost imperceptibly from day to day. In this lies the practicability of utilizing the upper part of the San Juan River for slack water navigation. The river itself favors this, by reason of its singularly small water-shed, other than that of and through the lake. On the one side, it is flanked by the Rio Frio, which runs in almost the opposite direction, and empties its waters into the lake, and on the other, by a very narrow water-shed above the mouth of the San Carlos, below which it is not designed to use the bed of the stream.

The River San Carlos, flowing from the mountains of Costa Rica, has the general features of inter-tropical mountain streams, and in striking contrast to the River San Juan in periods of floods, throws "back water" far above the Machuca Falls, 16 miles above, on the San Juan River. Below the mouth of the River San Carlos, from its floods and the vast quantity of debris transported by it, the utilization of the bed of the San Juan River, for slack water navigation, would be hopelessly impossible. This river, forming so marked a contrast to the San Juan, seems a reminder of the impracticability of utilizing the bed of any stream other than the San Juan, in that region, for slack-water navigation, or indeed of constructing a "thorough cut," by which is meant an inter-oceanic canal without locks. A canal so made must be the ultimate drainage of all the adjacent water-sheds, and must receive the debris of the mountain torrents, amounting, in the aggregate, to incalculable thousands of tons yearly.

A verification of the necessity of guarding against surface drainage falling into a canal is given on page 1 of the *Bulletin Decadaire* of the Suez Canal, of June 12th, 1878. In summing up expenditures for repairs is the following:—

"On the part of the canal in the vicinity of Suez, on the 25th of "October last, a violent storm produced a sudden and immense tem- "porary torrent, which, precipitating itself from the mountain of "Attaka, fell into the canal, carrying with it a very large mass of "solid material; this deposit, thanks to the activity of our employés, "was dredged very rapidly, so that the navigation was not retarded a "single day, but this unexpected dredging has added 30,000 francs "to the expenses of the canal."

The mean annual rainfall at Cairo, less than 100 miles from this locality, is one and three-tenths of an inch. At Aspinwall, the mean annual rainfall is 124.25 inches; during the month of November, 1870, there was a rainfall of 32.5 inches, twenty-five times the mean annual rainfall at Cairo. If the rainfall, as at Aspinwall, were the normal condition along the line of the Suez Canal, it seems altogether probable, from the damage and expenses caused by one shower, as quoted above, it would be a matter for serious consideration, whether it would not be economy to lock up sufficiently high, at least, to avoid the effects of the surface drainage. Then, if to the debris of a country denuded of wood were added the trunks, roots, and limbs of trees, it would be a still more serious difficulty to get rid of them also, even in the region of the Suez Canal, one of small water-sheds and easy slopes, as compared with any part of inter-tropical America.

Referring again to the line of the Nicaragua Canal, at and above the mouth of the River San Carlos, it leaves the bed of the San Juan

River on its left bank, and follows along the general course of the river for a distance of 28.1 miles. The plans provide ample culverts to discharge the heavy floods from the adjacent hills, with their debris, into the San Juan River, which bears it along until it reaches a point approximating to the ocean level, where the dynamic force is sufficiently weakened to allow it to rest. A failure to provide sufficiently for the surface drainage, in a country subject to heavy rainfalls, and withal of steep escarpments, would lead eventually either to the abandonment of the navigation of the canal, or to a vastly increased expenditure, to effect what could have been done at far less cost in the construction of the work.

At the distance of 28.1 miles below the mouth of the San Carlos River, the canal, as projected, takes a direct line to Greytown, a distance of 14 miles, passing through deep cuts, that it is now known can be avoided with a considerable decrease in the estimate for labor. When the location of the canal was made, the season was so far advanced that it did not allow the necessary examination to be made to improve this part of the canal line as located. Civil Engineer A. G. Menocal, U. S. Navy, has since made a sufficient examination to assure him of a more favorable location, with a probable decrease of the estimate for construction of $3,000,000.

A summary of distances and estimates of cost as given in the Report of Civil Engineer Menocal is as follows:—

Western Division.—From Port Brito to the Lake. Distance, 16.33 miles; estimated cost,	$21,680,777.00
Middle Division.—Lake Nicaragua. Distance, 56.50 miles; estimated cost,	715,658.00
Eastern Division.—From Lake to Greytown. Distance, 108.43 miles; estimated cost,	25,020,914.00
Construction of Greytown Harbor,	2,822,630.00
" Brito "	2,337,739.00
Total. Distance, 181.26 miles; cost,	$52,577,718.00

The expert who will carefully examine the items and estimated cost as shown by the Report and plans will assure himself that the object was to present fairly and in detail all of the work required, and at a fair estimate of cost.

It cannot be denied, however, that the estimates of cost and actual cost of construction have diverged so widely in very many great works, and notably with the Suez Canal, which had no uncertain element in construction except the drifting of the loose sands of the desert, that it would be a grave error not to recognize this fact. Had such been the case with the Suez Canal, and ample provision been made to carry on the work continuously, it is fair to presume that in the time gained for transit of vessels and consequent income, and the avoidance of the payment of interest on dormant capital, the actual cost of the canal might have been one-third of what the stock and obligations now represent.

A true economy, therefore, will be to consider the cost of the canal, including the interest on dormant capital, as double of the estimated cost of construction, in round numbers at $100,000,000.

The actual line of the transit, 181.3 miles, is far greater than the distances demanding labor; thus, to enable the lake to be navigated 56.5 miles involves labor only on a few thousand yards, at the inconsiderable cost of less than three-quarters of a million of dollars, and the river navigation, by slack-water, 63.02 miles, only $8,679,253, which includes four dams, three locks, three short canals around the dams, the diversion of the mouth of the San Carlos, and the necessary blasting and dredging; in short, the total cost over the distance named. Thus, actually 119.5 miles of transit is represented by an estimate of $9,394,911, nearly $1,000,000 less than the estimated cost of the feeder and its adjuncts of a canal *via* the Isthmus of Panama. The cost of the work falls really on the 58.23 miles requiring construction, the construction of seventeen lift-locks other than the three above named, and one tide-lock, and the construction of two harbors.

No less than twelve miles of the 58.23 referred to above, as entailing the heavy part of the expense, are so situated as to allow the work to be done by machine labor. Between Greytown and the point where the projected line of the Canal strikes the San Juan River, there is a large extent of alluvium, there being, in fact, several ridges of hard ground very favorably situated for the construction of locks, requiring but little cost for foundations.

The line of Canal being accurately marked, a rough tramway may be constructed on each bank, with foundations made of the felled trees and their roots, capable of sustaining and carrying along heavy derricks, and steam power to fell, cut up, lift, and deposit, first the trunks and limbs, and then pull out and put in place the stumps of trees, in the best manner to strengthen the embankments. Dredging machines can then follow, cutting their own channel, and deposit at once the soil to form the embankments. Should trunks of trees be found embedded, our Red River and other similar experience will enable the work still to proceed without material delay, or a very large increase in cost of labor.

This mode of procedure could be practiced also over nearly two miles on the west coast, leaving only about fourteen and a half miles on that section where the plow, scraper, and pickaxe would have to do the work.

The Government of Nicaragua is now engaged in blasting and removing rock from the channel of the San Juan River to improve its navigation for small vessels, employing the methods and appliances of Civil Engineer Menocal at a cost not exceeding two dollars and a half per cubic yard, which is one-half of the estimate made by him for that work; as the depth increases, doubtless the cost of excavation will also, but certainly not beyond the estimate on that part of the projected canal.

In relation to the formation of the harbors.—The Dutch as a people have done so much in constructing dykes and harbors, that their processes may be carefully considered if not adopted. Bamboos of large growth in any quantity are close at hand, which from their length and toughness may be an admirable substitute for small wood growths, so far as mattrasses are concerned. Abundant stone, sand, and a superior lime are near at hand on the San Juan, permitting a free choice of material in the formation of the harbor.

There is a well-established fact which I shall mention, as it seems to have a practical relation in connection with contours or facings of artificial harbors. This fact was observed by the late Rear-Admiral Davis and Professor De Sor, on Cape Cod, where the flood-tide "divides," one part setting north, the other south: *the debris of a vessel wrecked were invariably found in the direction the flood-tide set;* also, it was found that when coal vessels were wrecked on the south side of Nantucket Island, the coal was transported east, and then north, and deposited inside the hook at Great Point, this being, too, the course of the flood-tide.

There is no engineering difficulty in the construction of this work; with the exception of the uncertainty as to cost of the harbors, and the probability of finding buried trees in the excavation of the alluvium, there seems to be no likelihood of contingencies arising which would materially increase the cost of the construction above the estimates on any part of the projected line of the inter-oceanic ship canal. I shall have occasion, on referring to other proposed routes, to point out a marked difference in this respect.

The surveys and location of the Panama route for an inter-oceanic ship canal were also made by Com. E. P. Lull, assisted by Civil Engineer A. G. Menocal, U. S. N., and a sufficient number of young navy officers to carry on the work vigorously, which was begun in January, 1875. An actual location was made along its entire length, and calculations also of approximate cost on a common basis with those made of the Nicaragua route by the same officers. The estimates and all information published appear in the Appendix to the Report of the Secretary of the Navy, in 1875; the plans, etc., were made as fully as those of the Nicaragua survey, but owing to a failure of an appropriation by Congress for that purpose, have not been published.

The length of canalization required is 41.7 miles; a viaduct is necessary over the Chagres River, of 1900 feet in length, the surface-water in which would form the summit-level 123.75 feet above the mean (half) tides of the oceans. Twelve lift-locks on each side and one tide-lock on the Pacific side are required, twenty-five in all, being four in excess of the Nicaragua route. A feeder and adjuncts are required at an estimated cost of $10,366,959, which, as before stated, is $972,048 greater than the estimated cost of the whole distance improved and not canalized on the Nicaragua route, a distance of one hundred and nineteen and a half miles. In this connection, it will be remembered that the Nicaragua route requires no feeder.

The total estimated cost of the canal by the Isthmus of Panama is $94,511,360, in round numbers double that of Nicaragua.

The level of the highest water-mark apparent at the point proposed for the construction of the viaduct was found to be seventy-eight feet above mean (half) tide. As the ordinary elevation of water at the same point is only forty-two feet above the same level, this great rise, thirty-six feet, known to have taken place in six or eight hours, would cause serious apprehension at times for the safety of the viaduct.

The water supply that year, 1875, was supposed to be ample, and it was stated that the water was unusually low. In the month of April of this year, Civil Engineer Menocal, from personal observation of the Chagres River, regarded the water supply as inadequate; the volume of water was in fact reduced at that time to a mere rivulet. This deficiency of water may be regarded as of rare occurrence, and at a fixed period of the year, and could be ameliorated, if not obviated, by the construction of extensive, and no doubt expensive, reservoirs on the upper waters of the Chagres. The most serious defect of this route is shown in the Report of Commander Lull, as follows: "Large vessels, of eighteen feet draught "and over, of which, as I believe, there will be very few in the future, "would have to wait for a favorable stage of the tide, to enter or leave "the canal, making a delay, in extreme cases, of from five to six "hours. The enormous cost of giving a deeper channel we regard as "a worse evil than any small delays to ships in passing."

Commander Lull sums up the advantages and disadvantages of the Panama route as follows: "The advantages of this line are: an "ample water supply; an open cut, with but a moderate depth of "excavation; a comparatively short distance from sea to sea; fair har-"bors on either side; the proximity of a well-constructed railroad; the "established communication with the principal ports of the world; the "absence of high winds; and, in common with the whole isthmus, "the fertility of the soil and the salubrity of its climate during the "dry season.

"The disadvantages are: the large annual rainfall; the want of "material for construction purposes; the character of some of the "swamp lands in certain periods of the line; the amount of tunneling "required in the feeder; the necessity of a viaduct; the prevailing "calms of Panama Bay, causing tedious delays to sailing vessels; "and, finally, as compared with more northern lines, the greater dis-"tance of Panama from the west coast of the United States."

The want of material for construction purposes enumerated among the disadvantages would lead to heavy expenditures in cost and transportation, as compared with the Nicaragua route, where abundant material of all kinds is found convenient to, if not on, the projected line. Owing to the great prevalence of rain and the lack of material at hand, it is probable that the actual cost of construction would exceed fifty per cent. for a given amount of work as compared with the Nicaragua route.

The construction of the Panama Canal as located requires a "summit cut" of 4.81 miles in length and an average depth of cutting of 76 feet above the proposed water-surface of the canal, which would make a cut of 102 feet necessary to ensure the passage of vessels of 24 feet draught.

The liability to landslides in this deep cut would be very great, as is the case along the line of the Panama Railroad, necessitating a slope probably of two to one; and of course, in making the excavations, requiring the considerable expense of removing the material to a place of deposit. With all of these grave disadvantages, it may be admitted that the Panama route should be regarded as practicable, did not a better one exist. It is, at least, far superior in practicability to any line lying south of it.

In relation to other actual locations made for the construction of an inter-oceanic ship canal, what is known as the Atrato-Napipi route was surveyed by Commander T. O. Selfridge, U. S. N. More detailed and specific locations were made throughout its length by Lieut. Frederick Collins, U. S. N., in 1875.

A careful examination of their reports will reveal very great difficulties to be overcome, and the doubtful results practically attainable, on a common estimate of labor and cost of execution, with what relates to the two routes already summarized; namely, the Nicaragua route and the Panama route.

The reports and location of two routes, by Lieut. Wyse, of the French Navy, indicate the existence of the same grave difficulties in exaggerated dimensions that belong to the Atrato-Napipi route. Of the two, he prefers what may be designated the Tuyra-Tupisa-Tiati-Acanti route, which, by the employment of a tunnel, the length of which is still uncertain, is projected as a canal without locks; consequently the sea-level will represent the surface-level of the canal, except the additional elevation due to the surface-drainage which the canal must inexorably take.

I will state, briefly, the physical conditions and the methods proposed by Lieut. Wyse to overcome the very many grave difficulties which the projected line of ship canal presents.

The first section of the projected canal consists in deepening the channel-way, where necessary, of the Tuyra River, which falls into the Gulf of San Miguel, and through the improvement of which the river is intended to be made navigable to the commencement of the second section.

The second section is a cut from the Tuyra River, east in direction, to the junction of the Chucunaque and Tupisa rivers. This cut is intended to take the entire discharge of those two rivers, whose water-sheds, as far as can be ascertained from the maps, is approximately 1200 marine miles. Taking the estimated amount of excavation and length given, 16,200 metres (10.06 miles), gives a cross section of 705 metres as the mean, and the capacity to discharge, *with full banks*, the waters of the Chucunaque and Tupisa rivers. It will be shown, hereafter, how inadequate this prism would be to this discharge, with a velocity of water that would admit of navigation, during the period of floods at least, which is more than half of the year. In the absence of more specific information as to the volume of water in periods of floods of the Chucunaque, a quotation is made from page 135 of the Report of Commander Selfridge: "At one place, where the banks are about twenty feet high, we noticed "large trees, thirty inches in diameter, lodged at least thirty feet "above the ground, showing the great power and extent of the floods "during the rainy season."

The point at which this observation was made is on the Chucunaque, some twenty or twenty-five miles in a direct line above the mouth of the Tupisa, at which point the canal crosses the mouth of the Chucunaque. It is proposed across the mouth of the last-named river to erect a strong grillage, supported by abutments of masonry, to protect the canal from the floating debris. It is stated that this device will permit the waters to flow, and will arrest the trees, etc. It seems to me that the strong grillage would inevitably form a dam through the lodgment of drift-wood, and bank up the waters sufficiently either to cut channels around the grillage, or, if restrained sufficiently, that the waters would reach such a height as to carry away whatever opposed their free flow.

There are other grave difficulties in the proposed treatment of the second section which will be apparent to any one who will examine the Reports referred to. I will confine myself to expressing dissent to the supposition that the debris brought down by the Chucunaque need give no concern, as the strong tides would bear it away and deposit it where it would do no harm. Yet the reader of the Report will remark that dredging the bed of the river to a considerable extent on the first section is a part of the plan. In general, engineers will agree in the entire probability that dredging the bed of a running stream may be regarded as simply increasing the depth temporarily, and no longer in fact than the recurrence of floods such as formed the original deposit.

The third section is 11,400 metres (7.08 miles) in length. The projected work is to cut off the sinuosities of the Tupisa River, and deepen it sufficiently below the ocean-level to secure the passage of vessels of great draught. This is supposed to involve no difficulty; whatever obstacle is presumed to exist from the discharge of debris from ravines is to be met by the excavation of pockets at their mouths, and through the construction of grillages to protect the canal from drift-wood.

The total amount of material to be excavated and removed is stated to be 9,760,000 cubic metres, 500,000 of which is supposed to be rock. This mass is to be excavated and *transported somewhere*,—a most difficult matter in the execution of the work, when it is remembered that it includes excavating and removing half a million metres of rocks below the sea-level.

The fourth section, as projected, leaves the bed of the Tupisa River and extends to the entrance of the projected tunnel; it is 17,000 metres in length (10.56 miles), and has a mean depth of excavation of 33 metres (108.26 feet), including the part designed to be below the surface-level of the ocean. It is stated that the upper layer is vegetable detritus two or three metres thick, then a layer of six or seven metres of clay, overlying rock, easily excavated with the pickaxe to a mean depth of 23 metres (90.5 feet). It is assumed that this rock can be excavated and removed at $1.20 per cubic metre; this sum is one-fifth the amount estimated for rock excavation in the bed of the San Juan River, in Nicaragua, where no difficulty exists in depositing the material when excavated. It is obvious that all of this material, composed of vegetable mold, clay, and rock, has either to be hauled up and placed far back of the escarpment, to prevent its washing in again, as much as possible; or it has to be removed on the sections of the canal previously constructed, through the section from the point of excavation, and through the completed section, a mean distance of twelve miles; and even then, no dumping-ground exists nearer than the ocean, without at least rehandling the whole mass.

Page 95 of the first Report states that the bottom of the canal in sections 2, 3, and 4 will form an inclined plane equal to 4.75 inches to the statute mile. This inclined plane would, it is stated, make the tide from the Pacific almost nominal at the western end of the tunnel. Taking the distance from the beginning of the second section to the end of the fourth, and the proposed rise, and adding thereto the same incline from the head of Darien Harbor for the corresponding distance per chart, would make the total rise of the bottom five and six-tenths metres (18.37 feet); with this grade there seems to be no reason why the spring tides should not continue on through the tunnel into the Atlantic.

No depth of water is given for sections 3 and 4, but certainly it is designed that it shall be sufficient to float a vessel of heavy draught at full tide at least. If the tide is to cease at the entrance of the tunnel, it would seem that this could only be effected by a tide-lock (which is not proposed), or by increasing the incline or grade so that the bottom of the canal at the entrance to the tunnel would cut the plane of the high water at that point, or that the water in the canal below should be sufficiently shoal to retard the flow of the tide, in which cases, whether sufficient or insufficient for vessels to pass, the surface-drainage would represent the actual depth of the canal at this point at full tide.

The extreme tides at Chepigana are given by Commander Selfridge at about 24 feet, and by Lieut. Wyse at about 25 feet, with a very considerable difference in the height of the neap tides. These differences in height would be necessarily continuous above, as far as the deepened and straightened channel permitted, and the deepening and straightening of the channels of these streams sufficiently to permit navigation would certainly enable the tide to have a free flow.

The inclined plane of the excavation as proposed—4.75 inches per statute mile—would give the width and depth necessary for a ship canal, a current of not less than 2.2 miles per hour. The tidal action under normal conditions of water-flow would increase this current greatly, and when floods would occur we may well suppose that in earth excavations the scouring effect would be very great, destroying in a great degree the excavated grade, and depositing the material where it may or may not obstruct or entirely bar the passage through the canal to vessels of heavy draught.

This ship canal, as projected, presents the remarkable condition of inviting and receiving the surface-drainage, without the interposition of feeders, being itself the waste weir for a large superfice of mountainous country subject to extraordinary rainfalls. Imagine the effect on this canal-bed when a sudden rainfall of six or eight inches would make these mountain ravines actually roaring rivers, which in the dry season, owing to the rapid fall, are almost without running water.*

---

* In Washington, June 10, 1876, there was a rainfall of 2.03 inches in one hour. July 30th, of the same year, there was a rainfall of 4.12 inches in eight hours and a half. In October of this year there was a rainfall of 4.44 inches in twelve and a half hours, nearly all of which fell in nine and a half hours. The mean annual rainfall is 41.54 inches. In 1877, it was 52.59 inches. At Aspinwall, mean annual rainfall 124.44 inches; in 1872, 170.16 inches.

This method of location imposes the extraordinary and onerous condition of *having no natural dumping-ground* for enormous masses, such as the excavations proposed. The canal itself being the lowest level short of the depths of the ocean, renders it necessary either to rehandle the whole mass, or to transport the material to the ocean. In short, to make the excavation is a small part of the integral of cost, the transportation of the material being in fact the most expense.

The fifth section is a projected tunnel 36 metres high (118.1 feet), of which 10 metres (32.8 feet) are below the sea-level, 16 metres (52.49 feet) wide at the bottom, and 20 metres (65.6 feet) at the surface or ordinary, or rather *intended water-level*, and of still an uncertain length at the time of the publication of the second Report. The shorter length named is 9300 metres (5.78 miles), and the longer 18,500 metres (11.49 miles), almost double the first distance.

How this tunnel is to be excavated and walled ten metres (32.8 feet) below the sea-level is not stated; if not excavated and walled in water, it must be kept free from water by means of pumps. It is safe to say that the enginery of the world could not be placed to effect this object, during the construction of the work. The whole region tunneled through has a mean elevation of hundreds of feet, and peaks of more than two thousand. Beyond a doubt, a line of tunnel excavated 32.8 feet below the ocean-level would uncover sources of water of great power, sufficient to wash out whatever was within the tunnel as far as excavated.*

---

* The excavation of St. Gothard tunnel, now in progress, is the most recent development and illustration of what may be done in overcoming difficulties. The following information is from Johnson's Cyclopædia:—

Length of tunnel, 14,900 metres;. terminations Airolo and Göschenen, respectively 1145 and 1105 metres above the sea-level. The centre of the tunnel is to be 1154.4 metres above sea-level, *making a grade of one metre to the thousand* toward Airolo to get rid of the water; the other grade is 5.82 to the thousand.

On the high Alps the rain precipitation is small, as compared with the Isthmus, yet with a grade of 1·1000 for drainage. "The work has been frequently impeded by the "caving in of the rock, and by the irruption of the water from fissures in the strata." Imagine the increased impediment were the bottom of the tunnel 32.8 feet below the surface-level of the ocean.

On page 206 of the Report of the Secretary of the Navy for 1875 will be found the instructions to Commander Lull, in relation to the examination for tunneling between the Chepo River and the harbor of San Blas. Page 221 gives the deductions of Commander Lull, and a rough estimate of cost of tunnel.

This route would, in fact, require inconsiderable labor or difficulty except the tunnel, which would be, in length, little more than one-half that of the greater length named as possible by Lieutenant Wyse, on his projected route.

In the opinion of the ablest engineers to whom I have presented the question, it is impossible to make an approximate estimate of the cost of tunnel, as conditioned by Lieutenant Wyse, of the time necessary for its completion, or to state satisfactorily any known methods which could ensure the execution of the work.

The sixth section involves a length of 10,000 metres (6.21 miles), and a depth of cutting of 25 metres (83.33 feet), including an intended depth of 8.7 metres (28.54 feet) below the sea-level.

At this end of the canal it is proposed to construct a harbor which would not be less expensive than the one proposed at Brito, on the Nicaragua route.

Let us conceive that this stupendous work, from ocean to ocean, is completed, tunnel and all, as proposed. As shown by data, a rainfall of six inches, in as many hours, is not at all unusual in that region. Obviously, from the shorter distance to the sea, the readiest escape of the waters of the Tupisa Valley would be through the tunnel. We can form some idea of the relative size of its water-shed, with steep escarpments to its excavated bed; and also of the dimensions of the cross-section of the tunnel, which is the continuation of the artificial channel of the Tupisa towards the Atlantic.

Within an hour or two a perfect deluge would pour down the ravines from a thousand times the superfice of the continuation of the Tupisa channel; the rush of waters through the tunnel would be terrific,—sufficient, probably, to destroy almost the vestige of it.

With the exception of the surveys of Lieutenant Wyse, of the French Navy, above alluded to, all of the routes surveyed—indeed, I

A tunnel between the Chepo and San Blas Harbor could be constructed to free itself from water and have, probably, a sufficient water supply from the Chepo, rendering lockage necessary only sufficiently high to give drainage for the construction of the tunnel. Commander Lull says, "The line bears no comparison to either the Nicaragua "route or that of Panama as developed."

A distinguished officer of the Engineer Corps of our army informs me that the *head room* of the tunnel, as proposed (85.3 feet), would be insufficient; that in a bridge in our country 135 feet was regarded as barely sufficient.

The width at the water line as proposed (65.6 feet) is little more than the actual beam of the class of vessel that would usually be employed, say a beam of fifty feet, one-ninth the length of a vessel of 450 feet. This would leave less than eight feet on each side, provided that the vessel was pointed *absolutely fair*, which would rarely be the case; the slightest deviation, of course, would throw the bow of the vessel on one side and the stern on the other. It was not my intention to object to the size of the tunnel, from the fact that it could not, by any possibility, be constructed as a paying investment, even though it be admitted possible as an effort of the engineer, without reference to cost. If the tunnel is to be made, then the dimensions should undoubtedly be increased.

may say, all of the most practicable routes which exist—have been carefully examined by the Commission appointed by the President of the United States. To satisfy the wishes of the Commission, our Government executed the survey and location of an inter-oceanic ship canal *via* the Isthmus of Panama. The Commission was composed of the Chief of Bureau of Engineers, U. S. Army; the Superintendent of the U. S. Coast Survey; and the Chief of Bureau of Navigation, U. S. Navy. No difference of opinon existed in the Commission, as shown by their Report to the President, which indicated, in a marked degree, their opinion of the superiority of the Nicaragua route over all others. The recent surveys of Lieutenant Wyse have, in fact, served to confirm the justice of the conclusions of the Commission.

The question then at issue is, Can an inter-oceanic canal be constructed with advantage to its constructors, as well as to the advantage of the commerce of the world.? The points between which the trade would naturally pass through the canal are shown by the map.

1. Between the Atlantic coasts of Europe and America and the Pacific coasts of this continent.

2. Between the Atlantic coasts of Europe and America and Japan, Northern China, the Philippine and Sandwich Islands in the northern hemisphere, and the eastern coast of Australia, New Zealand, and numerous productive islands in the souther hemisphere.

The Chief of Bureau of Statistics has kindly furnished me with several tables, two of which relate to the tonnage of the trade that would naturally pass through this canal if constructed. Additional information from the same source and from Owen M. Long, Esq., U. S. Consul at Panama; also from Mr. P. M. McKeller, U. S. Vice-Consul at Valparaiso, indicate that not less than 3,000,000 of tonnage, British and our own, would pass through the canal yearly, and that the aggregate of the other commercial powers would be very large. I have to regret that my statistical information was received too late to collate it properly; its publication otherwise would simply lead to confusion. A glance, however, at the points between which the trade would pass through the canal cannot fail to be as satisfactory and convincing as the most elaborate and accurate tables.

TABLE C.

TABLE OF DISTANCES BETWEEN CERTAIN PORTS VIA CAPE HORN, AND VIA NICARAGUA CANAL.

	Distances via Cape Horn.	Via Canal.	Difference.
New York to Valparaiso	8,720	4,626	4,094
Liverpool "	9,100	7,326	1,774
New York to Callao	10,000	3,376	6,644
Liverpool ' "	10,400	6,026	4,374
New York to Honolulu	13,530	6,550	6,980
Liverpool "	13,780	9,200	4,580
New York to San Francisco	13,610	5,010	8,600
Liverpool "	13,665	7,600	6,065
New York to Yokohama	16,700	9,900	6,800
New York to Shanghai	14,500	10,300	4,200
New York to Hong Kong	17,420	11,550	5,870

I will not dilate upon the saving of distances shown by appended table, and other information contained in the admirable pamphlet of Professor J. E. Nourse, U. S. N., published in 1869. Its title is "The Maritime Canal of Suez, and comparison of its probable results "with a Ship Canal across this Continent." I commend it to all who feel an interest in this great question. An illustration may be given of the saving of distance from this port to San Francisco. Via Lake Nicaragua it is 5,010 miles, and through the Straits of Magellan 13,610—a saving of distance of 8600 miles.

A point worthy of mention is, that were it possible to use this canal between Atlantic ports and those of the Pacific without decreasing the distance of the voyage, the advantage of its construction would still be apparent, and would be appreciated by all intelligent navigators on account of the favorable winds that would be thereby obtained, on both outward and homeward voyages, through making very slight detours to effect that purpose, the canal route taking the place of the tempestuous seas of Cape Horn and its vicinity, and in their stead giving trade-winds in the region of the tropics, and westerly winds, and good weather usually, in the higher latitudes adjacent to that region.

A work of such magnitude as this inter-oceanic ship canal, so distant from the money-centres of the world, requiring a considerable time to complete it, even if prosecuted in the most vigorous manner, and intended to serve world-wide interests, could without doubt best be constructed on an international basis. The United States is relatively near the work, and has a twofold interest—the one to unite her coasts by convenient water transportation, the other of external commerce for both coasts, and especially to secure to the west coast a European market for its average crop of 20,000,000 sacks of wheat.

The great commercial nations may fail for a time to recognize the full importance of executing this great work, or the means that will best forward it, but sooner or later, probably in the near future, the inter-oceanic ship canal *via* Lake Nicaragua will no longer be a problem, but an accomplished fact.

The construction of this work on an international basis would mark an era in the world's progress; it would, in a remarkable degree, facilitate the commerce of the world between the most distant points, leaving it impossible thereafter to make any work which would compare with it in practical results.

Its construction internationally would be "a sign and token;" it would show that modern diplomacy appreciated the possibility of *obtaining mutual and common advantages*, and had discarded the fallacy of old that diplomacy was intended for and only useful in what might be truly called "'sharp practice." Peoples have arrived at that intelligence that the Government of a nation may, in its relation to another, rather seek *to discover and promote common interests* than hope to obtain and maintain mean advantages.

In this intelligent view, the Government and people of Nicaragua have hitherto endeavored to bring about the construction of the inter-oceanic ship canal, offering to our Government some time ago ample powers and concessions to construct it, with faculties based on the broad and common interests of the world's commerce.*

This enlightened and liberal offer was not responded to at once, probably because it was deemed proper in advance to ask the views and co-operation of such powers as might feel disposed to participate actively in the construction of the canal. It is believed that all of the great commercial nations have been invited by our Government to consider the subject, and to express fully and frankly their views in relation to the most advantageous manner of bringing about the construction of the work.

In my paper, read two years ago, I endeavored to show that our transcontinental railroad interests would doubtless be promoted

---

* It is proper to state that the same views were entertained by Mr. Peralta, the learned and excellent Minister from the Government of Costa Rica, whose territory is contiguous to, and forms a part of, the boundary with Nicaragua, along the San Juan River.

through the construction of an inter-oceanic ship canal, and also, as a whole, that it would probably be advantageous to the Suez Canal, forming as it would a supplementary chain for voyages around the world, in regions of good weather and aided by fair winds. I have reason to believe that this same opinion is entertained by those interested in that great work.

It seems pertinent on this occasion to acknowledge, that to the courage, devotion, and ability of cultured officers as leaders, to Lieutenant Wyse, of the French Navy, and to Commodore Shufeldt, Commanders Selfridge and Lull, and Lieutenant Collins, of our Navy, and to their able assistants and followers, we are indebted for so much positive, substantial information, sufficient, in fact, to a full appreciation of what can and what cannot be accomplished. It is impossible for any one having no personal knowledge of the Isthmus to appreciate the difficulty of making surveys in that region. It is not in the power of man to make physical conditions; all that he can hope to do is to recognize them, weigh them properly, profit by them if possible to do so, and let them alone if they do not afford an advantage. It was in this spirit that the recent United States surveys on the Isthmus have been prosecuted, and their official description made.

In the scope of a reading I have found it only possible to indicate the sources from whence information could be obtained. I shall furnish tables bearing upon the subject, and give to the Society the results of my personal conclusions, and, as far as possible, the reasons therefor.

I beg to return thanks to our Chief of Bureau of Statistics, Joseph Nimmo, Esq., to Owen M. Long, Esq., U. S. Consul at Panama, and to Mr. P. M. McKeller, Vice-Consul at Valparaiso, for their valuable assistance, and to this audience for their kind and patient hearing of what I have had to say—far short, I feel sure, of what this great subject would permit.

THE PRESENT ASPECTS OF THE INTER-OCEANIC SHIP CANAL QUESTION.

BY REAR-ADMIRAL DANIEL AMMEN, U. S. NAVY.

Read before the American Geographical Society, December 9, 1879.

[THE FOOT-NOTES WERE NOT PRESENTED TO THE SOCIETY, BEING RATHER IN THE CHARACTER OF ILLUSTRATION THAN ARGUMENT.]

I am indebted to a prominent member of your Society for the suggestion that I should continue the discussion of the inter-oceanic question under its auspices. I propose for your examination, "The present aspect of the American inter-oceanic canal question."

I shall not tire the patience of my hearers by an attempted history of it; those who are desirous to inform themselves as to what was known and what was asserted prior to 1866 can do so by reading a Report to Congress of Rear-Admiral Davis, U. S. N., published that year under the title of "Inter-oceanic Railroads and Canals," of course referring to this continent. It contains from pages 31 to 37 the authorities cited,—a large amount of valuable information necessarily interspersed with much that is wholly unreliable or only of partial value. The elimination of mere assertions and of errors has added vastly to the work of exploration and survey since that time. For an outline as to what has been done since then, I may refer you to my paper of October 31, 1876, and a second read November 15, 1878, before your Society.

The first paper was designed to show the error under which M. Drouillet, French engineer, labored, and those who regarded him as an authority, in a pamphlet issued in Paris in May, 1876, apparently with the approval of the Commercial branch of the Geographical Society. It was designed to show that there did not exist unknown routes comparable for the construction of a ship canal to those already known. The second paper was to show the feasibility of a ship canal *via* Nicaragua as a commercial question, and to do this it seemed to me necessary to establish its superiority over all known points, especially as to economy of construction and permanency of works, from less liability to the destructive effects of floods or other probable causes.

In the month of March last, when our Government thought proper to have our maps and plans published,—the results of close instrumental surveys on the Isthmus of Panama, and also those known as the Atrato-Napipi route, for presentation to the Congress called to assemble at Paris on the 15th of May,—it was supposed desirable that I should go abroad to present them, with such other surveys as had been recently made under its orders.

I suggested that I had been one of a Commission appointed by the President on the 13th of March, 1872, for the purpose of examining into and reporting upon the question of a ship canal across the continent, and that on the 7th of February, 1876, this Commission had made its reports as to locality, etc., which had been acceptable to the Government. It seemed to me, therefore, that some other person should be selected to attend the Congress. A second objection was that the selection of the canal route was eminently a question for the ablest engineers only, and those of the highest character, to settle. I urged further that Commander E. P. Lull, of our Navy, who had been engaged on the surveys of Caledonia Bay and south of it, and afterward had been chief of the parties making the surveys of the Nicaragua and Panama routes, could well take my place, as he had a rare judicial capacity which would enable him to present fairly such work as he had done in comparison with the surveys made under the direction of other officers.

Notwithstanding these representations made by me, I found that the Government preferred my going as a delegate, when, of course, I appreciated fully the honor, and made such suggestions as seemed necessary for a proper discharge of the duties which belonged to the position. It seemed to me important that the civil engineer who had been employed on both the Nicaragua and the Panama surveys, and who had performed his duties most satisfactorily to the officer conducting them, should go also, either as my assistant or as a delegate, to make the technical development of the routes. The suggestion was carried out, and I may add, that so far as I am aware, Civil Engineer Menocal performed his duties with entire satisfaction to our Government.

In presenting important information to the Congress,—the results of our Government surveys,—it seemed necessary that certain ideas connected therewith should be expressed; for that reason, what I shall hereafter call my address to the Congress was prepared, and submitted to the inspection of the Department of our Government under whose instructions I went abroad.

On arriving in Paris the day preceding the meeting of the Congress, Mr. Menocal and myself lost no time in paying our respects to M. Ferdinand de Lesseps, too widely and too favorably known to require further comment, unless I may add, that then and on all other occasions our intercourse was in all respects agreeable.

On the morning of May 15th, preceding the meeting of the Congress, I received a visit from M. Blanchet, an agreeable French gentleman, who had been to Nicaragua on two occasions to secure a concession for the construction of a ship canal, and indeed had one, which last March was rejected by their Senate. I was informed by him that the previous evening M. de Lesseps had either caused a meeting to be held or an agreement entered into that was acceptable to M. Blanchet and to the party in the Congress who would support the Panama project. My understanding of this was, that they had agreed to permit the discussion of the question to stand on its merits, and that whichever side lost the vantage-ground would receive some recompense which had been agreed upon in advance. Once, at least, during the sitting of the Congress, I received a visit from M. Blanchet, who seemed to be very much excited about the question, and to think I should be also. I told him that the decision of the Congress was no concern of mine; that my duties would be discharged by making a fair and full presentation of all of the information in the possession of our Government, which was, in fact, the object of my being in Paris. After the adjournment of the Congress, previous to my leaving, I received another visit from M. Blanchet, who informed me, or at least conveyed the impression, that his opponents had acted in bad faith with him.

On the meeting of the Congress on May 15th, an immediate organization was effected, M. de Lesseps, President, and five Vice-Presidents. The names of the members of the different Commissions were called, numbered as follow: 1st, Statistics; 2d, Economic and Commercial; 3d, Navigation; 4th, Technical; 5th, Ways and Means.

After naming the members of the Commissions, M. de Lesseps remarked in a jocular manner that our work was all cut out, that many of the delegates were anxious to get home, and that we could carry the work through *à l'Américaine*, which may be translated, *with a rush*. The full meeting of the Congress was then adjourned until the 19th, the Commissions to meet at 9 o'clock on the following morning.

Owing to the non-arrival of our heavy package of books, maps, etc., I was not able to present them the next day as desired, but did so on the morning of the 17th to the technical Commission. On the presentation, I stated that I would have my remarks printed in French and English. The following day copies were in the hands of the Secretary for distribution to such persons as wished them. The address was as follows:—

GENTLEMEN:—The Government of the United States has conferred the honor upon me of presenting for the consideration of this learned and distinguished body the different surveys recently executed by its order, a part of them published only within the past month. In their order, from the north to the south, they are as follows:—

1. The survey of the Isthmus of Tehuantepec, by R. T. Shufeldt, now Commodore, U. S. Navy.

2. The survey of what is known as the "Nicaragua route," an actual location of an inter-oceanic ship canal, with several tentative lines, by Commander E. P. Lull, U. S. Navy, assisted by Civil Engineer A. G. Menocal, U. S. Navy.

3. The survey of what is known as the "Panama route," an actual location of an inter-oceanic ship canal between Aspinwall and Panama, including feeder, etc.

4. The Report of the surveys made by Commander T. O. Selfridge, U. S. Navy, extending from the Gulf of San Blas on the Atlantic and the Bayamo or Chepo River on the Pacific coast to the mouth of the River Atrato on the Atlantic, to the Gulf of San Miguel on the Pacific coast, involving many tentative lines, and thence following up the River Atrato 150 miles, and from thence up the valley of the River Napipi, known as the Atrato-Napipi route, and terminating on the Pacific coast at Chiri-chiri.

5. An actual line of location for an inter-oceanic ship canal, of what is known as the Atrato-Napipi route, terminating as before at Chiri-chiri, by Lieut. Frederick Collins, U. S. Navy.

Maps, plans, and calculations for material and labor on a common basis of cost are made for the "Nicaragua," "Panama," and "Atrato-Napipi" routes as located, affording a ready means of finally considering the relative cost of executing the work on the several routes.

On the 13th March, 1872, the President of the United States appointed a Commission, whose duties were "to examine and consider "all surveys, plans, proposals, or suggestions of routes of communi- "cation by canal or water communications between the Atlantic and "Pacific oceans across, over, or near the Isthmus, connecting North "and South America, which have already been submitted or which "may hereafter be submitted to the President of the United States "during the pendency of this appointment, or which may be referred "to them by the President of the United States, and to report in "writing their conclusions and the result of such examination to the "President of the United States, with their opinion as to the possible "cost and practicability of each route or plan, and such other matter "in connection therewith as they may think proper and pertinent."

A final Report was made by this Commission on the 7th of February, 1876, copies of which are furnished for the consideration of this Congress. It was composed of the Chief of Engineers, U. S. Army, the Superintendent of the U. S. Coast Survey, and the Chief of Bureau of Navigation, U. S. Navy. It held its sittings at various times, and considered all of the information then existing, and concluded that the various surveys and reconnoissance extending over the wide region involved were sufficient to arrive at a conclusion, except in the region lying in the vicinity of the Panama Railroad; it therefore requested the Government to have a survey made and an actual line of location for an inter-oceanic ship canal on the best route found practicable in that region, which was done without delay. The Government, at the same time, thought it advisable to have a more thorough examination and actual location made along the entire length of what is known as the Atrato-Napipi route. After a careful study of these surveys, maps, plans, and estimates, in addition to the information which was previously before it, the Commission made its final Report, before alluded to.

In the consideration of a great work, such as the construction of a ship canal across the American continent, we may well suppose that its permanency should be regarded as important as the selection of the route itself, involving the least cost of construction with the minimum of problems of doubtful cost in the execution of the work. With these points assured, the question becomes fairly debatable whether the physical conditions are to be considered too formidable to admit of the execution of the work as a commercial or monetary question—in fact, whether a grand idea for the amelioration of the

SHIP CANAL QUESTION. 47

great commerce of the world can be put in execution, or perforce abandoned, through the existence of obstacles too formidable in their nature to admit of an endeavor to overcome them.

Should it be considered, after a careful and minute examination of the question, that a commercial or monetary success is practicable in the construction of an inter-oceanic ship canal, whatever error may obtain through the selection of an inferior route through a misapprehension of conditions of permanency, or of first cost of construction in the location of the ship canal would work a double injury, in the failure to yield a proper dividend, by reason of an unexpected and extraordinary cost in construction, or constant demands for heavy expenditures in the endeavor to keep the canal navigable, and in the probable imposition of tolls, which would tend to drive away or fail to secure a considerable part of the tonnage which should naturally pass through it, and which would make the ship canal appear rather as an obstructor than the promoter of a world-wide commerce. I feel sure that these considerations will have weight in the mind of our distinguished President, at whose call this assemblage has met, to whose genius and indomitable energy are due the inception and the completion of the Suez Canal.

I shall leave to my able associate, Civil Engineer A. G. Menocal, U. S. Navy, a minute presentation of the surveys upon which he was engaged; namely, what are known respectively as the "Nicaragua" and the "Panama" routes. His note-books and other data will show that the plans and estimates are based upon substantial and sufficient information.

There are certain comparative conditions affecting the execution of the work on the three different lines upon which we give maps, plans, and estimates, which it is important to bear in mind in the consideration of the subject of the construction of a ship canal.

NICARAGUA ROUTE.

The rainfall is comparatively small. Our observations at Lake Nicaragua, extending over one year, show an annual rainfall of 48 inches, or 1.22 metres. [More extended observations give a mean annual rain fall at Castillo 83 inches, and at Granada 55 inches]. There is a distinct dry season of between five and six months, when work in progress would not be delayed or injured, and but little interruption need be apprehended in the rainy season on that section of the canal between the Lake and the Pacific, as the rain generally falls at night, with occasional showers during the day.

There is abundant good stone, hydraulic and other lime, wood, and bamboos, which latter may be found very advantageous in the construction of harbors.

There is a considerable population, well disposed, and when they can have remunerative employment, fairly industrious. The country has an abundant cattle supply of good quality for food, and other productions which would furnish the main subsistence for laborers on the canal, with a convenient water transportation in general along the line of ship canal as located, and lake communication with an extensive population and fertile region. This water communication can be greatly increased by the construction of a six-foot canal to Lake Managua, at an inconsiderable cost, and when completed would make the supplies of all kinds superabundant. Between Lake Nicaragua and the Pacific, near the line of the projected canal, several passable roads exist, and whatever other roads might be required over this short distance could readily be made at inconsiderable cost.

There is an inexhaustible water supply in its lake of 2800 miles of superfice, which equalizes floods and makes the daily changes small in the discharge of the River San Juan, by which it debouches into the Caribbean Sea.

It has an excellent harbor on the Pacific coast at San Juan del Sur, convenient for anchorage as Brito itself would be if improved as a harbor, inasmuch as the vessel in transit would have time to regulate her steam and be pointed fair to enter the canal at any assigned time. This reduces the necessity of a harbor at Brito to simply securing a perfectly smooth entrance to the canal.

Lake Nicaragua affords every facility for an interchange of cargoes that may be desired.

The west coast and the valley of the Lake are, as compared with the eastern slope, comparatively healthy, and upon the eastern slope a considerable part of the labor can be done by means of dredging-machines.

The approaches to both entrances are superior in advantages to those of either of the two other routes with which the Nicaragua route is compared.

These considerations would seem to warrant the belief that cost of construction, including material, would be far less than upon either of the two other routes with which the Nicaragua route is compared, as will be more fully shown hereafter.

PANAMA ROUTE.

The mean annual rainfall at Aspinwall in a series of seven years is found to be 124.25 inches, or 3.15 metres. A dry season exists, but it is limited to two or three months, lessening the effective time for labor and of comparative healthfulness of the laborers employed, the wet being the sickly season.

No building material suitable is known in that region. The ties and railroad telegraph poles on the Panama Railroad are brought from Carthagena or elsewhere.

The population is inferior to Nicaragua, and the country less able to furnish subsistence for a large number of laborers.

By means of the railroad already constructed a canal under construction would have a convenient transportation at whatever cost might be agreed upon.

The cost of the feeder and adjuncts, as well as other disadvantages, notwithstanding the shortness of the line, as shown by maps, plans, and estimates, make a total of $94,511,360, as against those of the Nicaragua route of $65,722,137, *on a common basis of cost of material and labor*, when in Nicaragua the material is near at hand, and subsistence abundant, and on the Panama route, or in its region, there is no material for construction, inferior subsistence, and less favorable climatic conditions for labor, as before stated.

ATRATO-NAPIPI ROUTE.

Although the mean annual rainfall is not known, there is no doubt of the fact that it is largely in excess of the rainfall at Aspinwall, on the Panama route. There is only a nominal dry season, as at any time a precipitation of several inches is likely to occur, and actually does occur many times yearly during the so-called "dry season."

The building material supposed to be available is confined to wood.

The population is so scant as to be unable to furnish either assistance or subsistence for even an inconsiderable number of laborers.

The River Atrato would furnish transportation to the mouth of the River Napipi. Along the line of the projected canal the country is alternately rough and covered with swamps, so that great labor would be necessary to construct roads to secure even wagon transportation for subsistence and material for construction.

Under such conditions the projected feeders requisite would be made at great additional cost, as well as the projected tunnel and locks. In dimensions the projected tunnel is as follows: length, 5,633 metres; height, 35.96 metres; width, 18.29 metres.

On the Atlantic slope there are twelve projected locks of 3.14 metres lift, and on the Pacific slope ten of 4.54 metres lift, the summit-level being 43.59 metres above mean tide.

With the view of having a definite comparison, the estimates for material and labor, so far as they are identical, were made on a common basis with Nicaragua. The cost on this basis is given as $98,196,894; but it is quite apparent that with the lack of material convenient, and of subsistence and transportation, as well as the absence of a dry season, and above all, the impossibility of making even an approximate estimate of the cost of a tunnel under such conditions, that the actual cost of the execution of the work would be far in excess of the estimate.

The same physical conditions—the absence of a dry season, and a general lack of material for construction, except wood, and the lack of subsistence—were found to exist by all of our parties, at various times, on what is known properly as the Isthmus of Darien, and of all the region lying south of it.

The long period of time over which the surveys of the United States have been prosecuted, designed to elucidate the problem of an inter-oceanic ship canal, indicates a persistent interest in this subject. I am happy to add that the present Chief Magistrate and his cabinet are fully alive to the benefit to be derived from a full consideration of the construction of an inter-oceanic ship canal, now that further researches of the topography of that region no longer promise a commensurate reward.

The people of the United States will look with great interest upon the discussions and deliberations of this distinguished convocation, and to suggestions which indicate the means that may be adopted to secure a speedy commencement of the work of an American inter-oceanic ship canal on such a basis as should assure its uninterrupted prosecution and early completion. It would seem that this object could best be accomplished by making the work actually international, could a proper and satisfactory basis of co-operation be arrived at.

The people of the United States recognize the great amelioration and benefit that the commerce of the world would derive through the completion of this great work, and are not disposed to regard the consideration of this subject solely with reference to the degree in which the commerce and interests of the United States will be relatively benefited through its construction as compared with the advantages that may accrue to other commercial nations. Such a ship canal cannot fail to be a great and common benefit, and especially in opening a rapid and easy transit between the Atlantic coasts of Europe and America with the western coast of America, and by the speedy development of Australia. Regarding this inter-oceanic ship canal when constructed as the greatest possible artificial highway that can be constructed, conferring benefits on all nations and peoples, the people of the United States consider its construction as one of common interest, and the guarantee of its neutrality a duty in common to all nations.

The presentation of maps, plans, etc., was followed by a technical presentation of the Nicaragua route by Civil Engineer Menocal. Afterwards, in answer to inquiries, he gave the methods of proposed improvement of the harbor of Greytown and the regimen of the bar as observed by him during several recent visits to that locality. I may add here properly, that the able sub-commission subscribed to the efficacy of the proposed method and as well to the method proposed for constructing the dams across the San Juan. Several engineers of note, at that time not favorably disposed to the Nicaragua route, made many inquiries with the view of developing its difficulties and its inferiority, and became so well informed as to adopt it as the route offering relatively the fewest difficulties, and, in the end, certainty of execution. These engineers were found afterwards among those who abstained from voting.

On the second general meeting of the Congress, May 19th, Sir John Hawkshaw, of England, whose reputation as an hydraulic engineer is second to none, was present. The afternoon was taken up in a desultory discussion of the Panama route by Lieutenants Wyse and Reclus, of the French Navy. A considerable part of the discourse was directed to the Nicaragua route, which was not under discussion. The data upon which their plans were constructed were quite insufficient. The cause of the anxiety of Lieutenant Wyse, when in the United States, two months before, to obtain tracings of our maps and plans became at once apparent. They were not furnished him because it was considered improper to give them publicity abroad in advance of their publication at home.

It will be remembered that previous to last autumn, after making an examination of the valleys of the streams falling into the Bay of San Miguel in 1876-77, and visiting that region the following season,

Lieutenant Wyse made plans and estimates for two routes, the one preferred by him called the Tuyra-Tupisa route, which by his Report was supposed equal, or nearly so, to any that had been developed through our surveys. This route seemed to me hopeless from the existence of the gravest difficulties, some of which I mentioned in my paper of November last. It seems from what I shall presently quote, that Lieutenant Wyse had the frankness to inform the Society for which he was acting, that in his view a ship canal across that region was impossible. He did not present it at all in the Congress, but took up the Panama route on whatever information he had, and developed it for a ship canal à *niveau*, which certainly was a step in the right direction. It may be said without dispute that for a canal at the ocean-level, the Panama route is far preferable to any other. The *possibility* of it must be considered simply in a commercial sense, as a canal, if made at all, must have that condition.

The following day, May 20th, Civil Engineer Menocal was invited to explain the plans and estimates of the Panama route, and was so interrupted by questions that Sir John Hawkshaw suggested allowing him to proceed and making questions afterwards.

He stated that when Commander Lull and party began the survey of the Panama route, there was no pre-occupation as to what height above the sea, if any, would be selected as the summit-level. They found at Matachin that the floods of the river passed some five or six feet over the railroad track, and that at low water the surface of the stream was forty-two feet above the ocean-level. In considering the question, it became apparent that if the ocean-level were adopted, an excavation would be necessary, making the normal surface of the proposed canal forty-two feet below the present low water, which would then make a small cascade, and in periods of floods would be transformed into a cataract of one hundred and sixty-one thousand cubic feet per second, from a height of nearly seventy-eight feet, the decrease being due to the increase of the velocity of the water as it approached the precipice, and also to the head of the water above the ocean-level after falling, which would give a corresponding velocity on its course to the sea.

It was apparent that either this great volume of water must be received into the canal from an elevation which would make the effect destructive, or that it would be necessary to lock up so as to permit the floods to pass beneath the aqueduct. This would bring the surface-level of the water in it to an elevation of one hundred and twenty-four feet above the sea-level. This was found to entail the construction of a feeder, with its adjuncts, at a cost of $9,942,727, with either a doubtful or a scant supply during a portion of the seasons of unusual draught. On concluding, Mr. Menocal stated his willingness to answer questions without eliciting any.

On the 21st another general session was held. Sir John Hawkshaw gave his opinion on the Panama route, as follows:—

"With regard to the question whether the canal should be constructed with or without locks, the following points occur to me:—

"If the canal is to be without locks its normal surface-level would be that of the sea, and its bottom-level, say eight metres lower. This being the case, the canal would receive and must provide for the whole drainage of the district it traversed.

"Therefore it would be necessary to ascertain the volume of water that would drain into the canal before it would be possible even to determine the sectional area of the canal.

"If the canal have a less surface-fall than the river, as it would have, it must have a larger sectional area to discharge the same volume of water.

"The average section of the river in a flood at Mameï was ascertained by Mr. Reclus to be 1310 square metres. This would require a canal, if it were eight metres deep, to be 160 metres wide.

"The waters of the Chagres would have a tendency to flow toward the Pacific, that is, through the tunnel, as the distance is less and the fall greater than to the Atlantic.

"It seems to me that the dimensions of the tunnel, if it has to serve for both the river and canal, would be too small. Mr Menocal's estimate of the volume of the Chagres in time of flood would much more than fill the tunnel; and in any case the whole section of the tunnel is only half that of the river in time of flood, as given by Mr. Reclus.

"During the construction of a canal at the sea-level difficulties would arise in providing for the drainage, which would affect both time of execution and cost to an extent that could hardly be ascertained in advance.

"If, from such considerations as the foregoing, it should be concluded that the canal should be so constructed as to retain the rivers for natural drainage, then recourse will have to be had to locks.

In that event there can be no difficulty, in my opinion, in carrying on the traffic with locks properly constructed, provided there is an ample water supply, which would be a *sine qua non.**"

It will be observed that Sir John expressed the axioms heretofore acknowledged by able engineers: to avoid surface-drainage, and to have an abundant water supply.

After reading his opinion, he remarked that a residence of two or more years in inter-tropical America had given him a knowledge of how these showers behave, without which he might think differently. In a conversation with him before he left Paris, after two days' attendance at the Congress, he expressed the opinion that the canal could not be excavated *à niveau*, and if it were, that it would be filled up with trees and silt.

A pamphlet by V. Dauzats, Chief Engineer of the Suez Canal, compares that canal with the various routes proposed across this continent. He quotes at length from my last paper read before this Society, showing the marked contrast of physical conditions, the region of the Suez Canal having a mean annual rainfall of less than two inches, whilst the region of the Panama Canal has a rainfall of one hundred and twenty-four inches. His deduction is, that surface-drainage falling into a canal has a scouring effect which is beneficial, whilst the abrasion of the banks of a canal is far more destructive. Were it not too great a tax on your patience, I would point out the fallacy of such an argument. It is assumed that when a river as the Chagres is dredged it will change its regimen. This deduction is necessary to a supposition that a canal *à niveau* at Panama is possible.

On the afternoon of the 19th the Technical Commission was divided: one part to report upon the practicability of locks as presented on the Nicaragua route, the other to consider the question of making tunnels for navigation. There was confusion and violent action, I was informed, on the part of Lieutenant Wyse, growing out of his opposition to Mr. Menocal being put on the sub-commission on locks. Mr. Menocal very properly asked to be excused.

---

* Through the "Proceedings of the Royal Geographical Society of London," page 608, of September last, we learn that at a meeting of the Geographical Society of Paris, M. de Lesseps said that the plans of Wyse and Reclus were undergoing modifications in the substitution of an open cut for a tunnel, and providing for a new bed for the Chagres River, the latter not considered necessary in the Congress by M. Dauzats and others. It is highly complimentary to Sir John Hawkshaw that after the adjournment of the Congress his ideas have more force than when presented. The increased estimates, especially for the latter, should be simply enormous.

The Report as to locks was that they could be made to serve their purpose. The calculations for a tunnel were made for construction on a dry foundation; it was stated there were no elements of calculation for building a tunnel below the sea-level as the plans demanded.

During the sitting of the Congress I found myself frequently obliged to dissent from the propositions of Commander Selfridge, U. S. Navy, who, strangely enough, was found in the Congress without being named by our Government.

This officer had been the chief of large parties who were engaged during the seasons of 1870, 1871, and 1873, in examining the coasts lying south and east of the Panama route, at San Blas, Caledonia Bay, the streams flowing from the flanks of the mountains adjacent to the Bay of San Miguel, and of the counter-slopes falling into the Atlantic, also in making an examination of the Atrato-Napipi route for a ship canal, which will be found in his Report to the Secretary of the Navy of June 12, 1873.

I refer the curious reader to pages from 66 to 70 inclusive, and to map VIII., illustrative of the Atrato-Napipi route as developed by Commander Selfridge. Nobody reading this Report and referring to the map would suppose for an instant that the greater part of it was purely imaginary, the ground lying between the rivers Atrato and the Doguado never having been passed over by Commander Selfridge or any of his party. It is delineated as an inclined plane, locks located, and sections of elevations given in figures! Between this fanciful presentation and the profiles made by Lieutenant Collins, U. S. Navy, there is a very wide difference. I quote from page 7 of my Report:—

"Commander Selfridge then said that the remarks made by Sir John Hawkshaw in relation to the Chagres River were not applicable to the Atrato-Napipi route, and endeavored to enter into a further discussion of its merits. I stated that I would suggest the advantage of discussing the carefully prepared plans of Lieutenant Collins along the lines of actual location, which were the best that could be found in months of labor, instead of lines drawn at will by Commander Selfridge, involving uncertainty of execution and an entire absence of elements of calculation, as every engineer would recognize."

This was one of several occasions that I had to suggest the advantage of discussing facts instead of indulging in fancies calculated to deceive the credulous and the unwary, and absolutely a waste of time in discussing.

The proceedings of the General Congress on the 23d, and in the Technical Commission on the 26th, are so significant that I shall append them without omissions. By reference to the Appendix, it will be seen that the partial quotations which I shall use do not present a perverted meaning. I will submit the question to every reader of the Appendix, whether, free from any comment, it is not patent that the Congress was not called to decide upon the best route for an inter-oceanic ship canal, but upon what was *possible via* Panama.

M. DE LESSEPS.—"That which struck us the most, is the enthusiasm of the United States of America in favor of the establishment of a canal at Panama."

We may ask with surprise, when and where was this enthusiasm manifested? I saw nothing of it; so far as my expression is concerned, it requires only very ordinary perceptions to accredit it as something more than that of an individual, inasmuch as I had been sent there by my Government.

I again quote M. de Lesseps.—" Lieutenant Wyse and his companions have rendered us an account of the mission that they undertook. Seven of them set out, four are dead in those wilds where one is only able to effect a passage with a hatchet in the hand. They have then returned, and have had the honesty to declare to us that in their view a canal was impossible in the regions that they had returned from exploring." This seems sufficient to dispose of the historical sketch of M. Hertz, given on page 10 of the Proceedings, as follows: "The French committee of study for the inter-oceanic canal [in consequence of the completion of the surveys alluded to by M. de Lesseps] thus found itself able to submit to an International Canal Congress a collection [of information] upon which it would be able to pronounce intelligently. It is known with what alacrity the most learned men from all countries have responded to the call."

To show the sufficiency of our information previous to these surveys of Lieutenant Wyse was the object of my paper read October, 1876, in reply to a pamphlet of M. Drouillet, who came to this country to obtain assistance in making further surveys. The closing paragraph of my paper was as follows: "I may add as a personal conviction, that however long and seriously the search may be continued for " results " by surveys, nothing can be or will be developed so advantageous as that which the surveys of our Government present for your consideration." Lieutenant Wyse's surveys undoubtedly destroyed pre-occupations in Paris, and so far was useful to them, which they might have effected at less cost by a more thorough examination of the work that had been done by our Government.

Notwithstanding what M. de Lesseps said respecting the assertions of Lieutenant Wyse as to the impossibility of a canal in that region, we find in Lieutenant Wyse's last Report a tabulated statement of routes, among which is the Tuyra-Tupisa, at an estimated cost of 600,000,000 francs.

I quote again M. de Lesseps.—" I have consulted M. Lavalley, and he has replied that it [would be] decided for a canal *à niveau* that it was a public sentiment. I will permit myself to sustain that opinion." Again, M. Lavalley has studied the question of a tunnel; he believes it certainly possible. He says " it is only a question of cost."

This Society will be surprised to find, on reading all that M. de Lesseps has justly said of the high qualities of M. Lavalley as given in the Appendix, that when the resolution was voted on, he, as also some other distinguished engineers of the French Society, were designedly absent. To the fact that these eminent engineers have not given the sanction of their names to what by others was regarded as *possible* in engineering is probably due the discredit shown to the decision of the Congress.

I quote again 'M. de Lesseps.—" In my belief we should not make a canal with locks at Panama, but a canal *à niveau*; that is, I believe, the opinion of the public, of which I am the organ at this moment."

Here we see, that instead of studying the question as an engineer, and in its economic conditions relatively with other routes, M. de Lesseps pronounces himself to be the organ of what he believes to be public opinion. Happily for the public, its supposed demand could not swerve M. Lavalley and others of great reputation.

I call attention to the remarks of M. Peralta as given in the Appendix. This learned and able minister of Costa Rica to our Government is well known to many of you personally. His suggestions were not to be considered; M. de Lesseps wished nothing more embodied in the resolution than whether a canal *à niveau via* Panama was possible. The resolution was passed as he desired, by such a vote as to call forth an expression of his satisfaction,—this, too, supported by the demands of public opinion, as he stated, and yet he is not happy.

I again quote M. de Lesseps.—" Since forty years I have studied the question [of the Suez Canal], I have always understood that for a profit it is necessary to receive at the least 10 francs per ton; one can perfectly well make the American Canal pay double that amount, whatever project may be brought about. These [are] considerations that one is very glad to know for the future."

The humanitarian idea so nicely held out, and especially supported, by M. Simonin, is dropped. There remains alone the idea of constructing a canal without reference to whether it is on the best location, but certainly on the line where the concessionists are entitled to receive by the terms of the concession 10 per cent. of the stock issued.

The Report of the Commission on Statistics of the Congress gives the tonnage likely to pass through the canal as follows:—

That of the United States,	2,000,000	tons.
" Great Britain,	1,050,000	"
" France,	356,000	"
" all other Powers,	356,000	"

In the *Bulletin du Canal Inter-Oceanique* of October 1st, published in Paris in the building of the Suez Canal, there is an article of some length entitled "*via* Nicaragua," in which is set forth in varied terms the egotism of the American Commission on the interoceanic canal question as shown in their Report to our Government, and also the same quality shown by our official delegates to the Paris Congress.

If this egotism was shown as is supposed in the Report, it was simply in the endeavor to promote the public interests in the most economic manner. The narrowness of the views of the Commission is supposed to be shown in recommending lockage for vessels of only four hundred feet in length and a beam much greater proportionately than that given vessels at this time. Without having the time or patience to look up the French steamers, I will venture the opinion that all of them longer than four hundred feet could be counted on the fingers of one hand.

The egotism of Mr. Menocal and myself at the Paris Congress, so far as I am capable of judging, was confined to a fair presentation of all of the information in the possession of our Government, and perhaps feeling no very lively interest in what the Congress would *decide*— not *determine*—which belongs to nature, and to the keen appreciation of moneyed interests as to what will and what will not pay. After the adjournment of the Congress an engineer very much in the confidence of M. de Lesseps said to me, "Now that the matter was settled, what amount of money might be counted on in America to promote the

SHIP CANAL QUESTION. 59

enterprise?" I replied that in my opinion they would not get a dollar. Evidently in my egotism I was wrong: to what extent will only be known when the *Bulletin* devoted to the canal interests publishes the amounts subscribed in France and elsewhere for the construction of the canal à *niveau*. Without assuming to speak for the public, I feel sure that such a statement would be read with interest.

Looking at the table just read of the tonnage of the different nations, we see the egotisms [interests] of all of them in form and substance. In the matter of the canal, the interests of the United States now are practically double those of Great Britain, and will become relatively greater proportionate to the increase of populations; those of France are, roughly, one-third of Great Britain, and yet if the word egotism is a proper substitution for the word interests, she has as much as all the rest of mankind.

In an interview given in the New York *World* of October 9th, M. de Lesseps is reported to have said, "If I may say so, I do not think the Americans are very clear-sighted on this matter. They are of the Anglo-Saxon race, and it is to some extent a question of race. The Anglo-Saxon race is unequalled for its power of dealing with the circumstances immediately before it, but I do not think it sees very far into the future. The Latin race has a somewhat wider intellectual horizon." He regards the Anglo-Saxon race as eminently practical; and without being of that race, I can well believe him. Granting his foreseeing power, may we not ask the probable number of Anglo-Saxons on this continent at the end of this century, and at that time, also, of those inhabiting Australia and the Pacific Islands? Awaiting his reply, may we not, without egotism, assume it to be, roughly, one hundred millions of people?

We can leave to M. de Lesseps, with his long view, the contemplation of the end of the next century, the period A. D. 2000. Still, even to our obscured vision, there seems a mighty multitude of men; shall we give it shape in supposing it to number at least 300,000,000?

Dropping the consideration of humanitarian ideas so unhappily dispelled, and looking at it as a plain business matter, could we not submit the question to the citizens of the two Powers first named, whether it would not be worth while to consider the construction of a canal on a *commercial basis*, and with reference to a careful examination of all of the points involved, and if found practicable in that view, do the work, and if otherwise, develop through the United States and the Canadas such additional railroads as would ameliorate the commerce of which they are so largely the factors?

After considering the proceedings of the 23d, in the general session, and a part of the proceedings of the Technical Commission of the 26th of May, as given in the Appendix, we can proceed to consider the vote more intelligently. A resolution was introduced to conform to the wishes, as expressed, of M. de Lesseps. It is as follows:—

"Le Congrés estime que le percement d'un canal inter-océanique à niveau constant, si désirable dans l'intérét du commerce et de la navigation est possible, et que le canal maritime pour repondre aux facilités indispensables d'acces et d'utilization qui doit offrir avant tout un passage de ce genre devra être dirige du Golfe de Simon à la baie de Panama."

Which I translate in these terms:—

"The Congress considers that the piercing of an inter-oceanic canal at a constant level, so desirable in the interests of commerce and navigation, is possible, and that a maritime canal to respond to an indispensable facility of access and utilization which a work of this kind should offer, should be located between the Gulf of Simon and the Bay of Panama."

The official vote as given in the proceedings is as follows: Abstentions, 12; against the resolution, 8; in favor of it, 78. The most significant figure is omitted. As counted up on the record, 36 were *absent*, among whom were a considerable number of engineers of note, and perhaps half a dozen delegates who were not in attendance during the session.

Had it not been that the expression of my abstention from voting was regarded as an "enigma," which has been *solved* in the *Bulletin* of October 1st, I would not have alluded to it. I abstained from voting on the ground that "only able engineers can form an opinion, after careful study, of what is actually possible, and what is relatively economical in the construction of a ship canal." I feel sure that it will excite a smile among us to suppose this in any degree enigmatical, and may recall the ideas so ludicrously shown in the comedy of the Irish Ambassador.

In relation to the vote and to the delegates, a pamphlet published in Paris, titled "Panama, 400,000,000 *a l'eau,*" gives the following: "Let it be remarked that one-half the members of the Congress were French; they had been chosen by the organizers of that assembly; 34 members belonged to the Geographical or the Commercial Geographical Society of Paris. What was their competency to decide between a canal with locks or on a sea-level? 14

other members were engineers or assistants of some sort on the Suez Canal. What was their impartiality to decide between M. de Lesseps and others? And among the others, if one takes count of personal friendships and of prestige exercised by a great name, how many more will remain?"

No one will deny that among the French delegates to the Congress were men eminent in every branch of engineering science, and others of the highest character as men of science. The same may properly be said of the foreign delegates. They were men of character and special attainments, usually having relation to the subjects that would concern a canal, if not its construction.

As regards the engineers of Holland and Belgium especially, where the land is so flat and the rainfall so small, their practical experience of a head of water would be confined almost to tidal action. However able they may be, they had not, so far as I know, that practical experience of inter-tropical America that made Sir John Hawkshaw so competent an authority.

Engineers in other branches would naturally adopt the opinions of the hydraulic engineers, and as far as their conscience would permit, be disposed to support the opinions and wishes of M. de Lesseps, especially if expressed emphatically, as found in the Appendix. They would say, very properly, the *execution* of the work was for M. de Lesseps, and not at all their affair. He had asked them to say that the canal *à niveau* was *possible*, and they had obligingly done so. He did not think it worth while to ask his honorable *confreres* if they thought Panama the best canal route; indeed, it would not have been prudent to do so, as he had determined that the canal should be built at Panama *à niveau*. As expressed by an engineer very much in his confidence: " If they found it possible, the first thing was to get the money, the next was to build the canal in the best manner that they found possible." Even a great general needs "the sinews of war." The public who made the demand through M. de Lesseps to have a canal *à niveau* should not desert him so cruelly. He has met them fully half way, in reducing the cost of construction one-half, as given by the Congress, and in still further shortening the time for the construction of the work, as given in his provincial tours, beyond that assigned by him in the Commission on the 26th, as shown in the Appendix.*

---

* After the Congress, comes naturally a period of rejoicing: of tours in the provinces, and of presentation to the Academy of Sciences,—the grand event, and the success of the Congress. I trust you have all read what he said on that occasion. I will only quote

I do not propose to discuss the terms of the concession fully, as found on page 281 of the Report of Lieutenant Wyse. I will point out some features that seem to me objectionable in the extreme.

The Canal Company agrees to transport gratuitously all persons in the civil and military service of Colombia, their baggage, arms, and ammunition, and if the company is not provided with vessels suitable for their transportation, to pay their passages and for the transportation of armaments and ammunition.

---

the closing paragraphs of his remarks : " I will add, as an event that seemed to me the most significant, finding myself yesterday at Nanterre, coming out of the municipal hall, upon the occasion of the annual fete of the Rosary, I was accosted by a group of peasants. One of them, as spokesman, said to me, ' When will you open the subscriptions to the American canal? We are with you.'

"By that voice of the people it seemed to me that I heard the *vox Dei*, and immediately took the resolution not to delay making a call in all countries for 400,000,000 francs."

That was an aspect of the canal question to M. de Lesseps and to the peasants,—a case, we may call it, of mutual self-deception. The illusions of fancy on the one hand were of a host of peasants bearing numberless stockings to be emptied of their gold and silver—a veritable bonanza! On the other hand, the simple peasants gazed with rapture upon what seemed to them a veritable Aladdin, who had happily presented himself to their mortal eyes, and taken upon himself to make them all rich, and spare them for the future from the toils of life. Beautiful, touching picture! that rested on the mental vision, then faded away into the cold, sombre, almost dismal realities of life. One thing alone can console in some manner: neither party is the poorer for having met the other.

The subscriptions proposed by M. de Lesseps for the inchoate Panama Canal Company were 800,000 shares of 500 francs each, amounting to 400,000,000 francs. It was intended afterwards to put a loan on the market for 200,000,000 francs, stock and loan amounting to one-half what the Congress estimated the canal *à niveau* would cost, without considering the cost of diverting the Chagres or constructing the tunnel on proper foundations 26 feet below the sea-level. The public has not been informed, so far as I know, as to the methods employed to so lessen the cost of the work.

Due notice was given that the books for subscriptions would be formally opened on the 6th and 7th of August last in all of the great commercial cities of Europe and America; intimation was given that before the formal opening a convenient back-door would be open for the appreciative and prudent to enter and subscribe, thus avoiding the annoyance and loss which might arise from not receiving all of the shares subscribed for after the books were formerly opened, but only a proper apportionment in view of the large excess of the subscriptions. The eventful days came and passed; although the object of the existence of the *Bulletin Inter-Oceanique* is said to be to inform the public on all matters pertaining to the canal, we are in ignorance of the actual amount of the subscriptions, and only know, as it were, that in the opinion of the President they were insufficient to warrant the immediate commencement of the work. In the absence of authentic information, we are indebted to other sources for the estimate that the amount of stock taken was a little in excess of one per cent. of the estimated cost of the canal.

The Government of Colombia is to receive semi-annually 5 per cent. of the gross receipts of the Company for the first twenty-five years; 6 per cent. for the second twenty-five years; 7 per cent. for the third twenty-five years; and for the remaining twenty-four years, 8 per cent.; at the end of which time the canal reverts to the Colombia Government.

The Company is "authorized to reserve 10 per cent. of the shares for the benefit of the founders and aiders of the enterprise." The only hope for a stockholder would seem to be in the extraordinary impost of ten francs per cubic metre, not on the gross tonnage or weight of vessel and cargo, but upon the cubical contents of a parallelopiped represented in the length, breadth, and draught of the vessel! Lieutenant Wyse supposed that this measurement might amount to 30 francs per ton, which, if imposed on ordinary cargoes at ordinary prices, as wheat, would make a voyage from San Francisco around Cape Horn preferable in economy.

Looking at the terms of the "concession," as it is called, and the whole matter from beginning to end, the wonder is that the subscriptions were so large rather than that they were so insignificant for the purpose of constructing a canal.

When Lieutenant Wyse was before a Commission in the Congress, he was questioned as to the Panama Railroad and its franchises, and replied that he had made a satisfactory arrangement by which the Canal Company would gain two millions of francs yearly, but gave no further explanation as to the arrangement.*

---

* I have been permitted to examine the articles of agreement of the "Civil International Society for the construction of the Inter-Oceanic Canal across the Isthmus of Darien," signed August 19, 1876, and also of the proceedings of its members in the partition of the *venture* in general assembly in Paris, June 10 and 17, 1879.

These interesting and instructive documents are not intended for the reading of the public; it would not add to the value of the shares or to the reputation of the concern. A single copy coming to the knowledge of moneyed interests would seem sufficient cause why the canal *à niveau* would receive no support.

A "concession" obtained by M. Gogorza from the Colombian Government had sufficient value in the eyes of General Türr and Lieutenant Wyse to admit of purchase and of initiating "surveys," and additional solicitations for amendments to the concession. In furtherance of additional surveys, M. Drouillet, French engineer, appeared in New York in August, 1876, and asked the co-operation of our learned societies to aid in a "serious attempt" to explore the Isthmus in behalf of science, and, doubtless, of this learned and, may we say, disinterested association? As an engineer, he declared that for five years he had endeavored to study the problem, but could not, as the information was quite contradictory. My paper read October 31, 1876, as before stated, was de-

So far, in general, we have been regarding the aspects of the interoceanic canal question from other points of view than our own, with occasional objections or remarks thereon. Let us now look at the question from our point of view. After the adjournment of the Congress, it seemed to me that its high authority, and that of M. de Lesseps, who does not, in the interview, overstate the confidence with which he has been regarded in France, would deprive many unfortunate peasants of their hard earnings. What kind angel protected them God only knows! So far as the English and our countrymen

---

signed to show that the *information* was sufficient and not contradictory, the confusion in his mind arising only from a want of comprehension of what was reliable and what was apocryphal.

The initiatory Society made "surveys" the value of which at most was to show that those which preceded them were sufficient and reliable within the limits claimed. At the time his "surveys" were in progress, Lieutenant Wyse, in 1876-77 and the following season, visited Bogota, securing in the end a concession which seems to me to afford no prospect of remunerating the constructors of the canal, even though that route was far more favorable than appears to those who made our surveys.

Then followed the calling of the Congress by M. de Lesseps, which gave authority to this scheme. A few days after the adjournment of the "Congress" there was a wrangle over the partition of the spoils, and, as usual on such occasions, a want of agreement, and finally differences and recriminations. In the discussion, which M. de Lesseps insisted should be in writing, he plainly informed his associates of his claims, growing out of his calling the "Congress" and the increased value of the shares due to its decision resulting from the confidence which his support inspired, without which the project might have failed. As a participator in the Congress, I can say with all frankness that I do not think that he overstated the value of his services. I do not suppose that he would be willing to state this as a compliment to the many able and excellent gentlemen who, in addition to other persons, gave him a support in the "Congress."

Referring again to this enterprising Society and M. de Lesseps, who wished to become the purchaser of their acquired "rights" of concession, and by bargain with the Panama Railroad Company, fifteen millions of francs for the former were demanded, after a considerable rebate, through the process of "dickering." To all this M. de Lesseps replied resolutely, *ten millions*—not a sou more! He would have no joint stock ownership or direction, which would seem to us more emphatic than complimentary to his former associates. This was passed over, however, as, after discussion among themselves, it was determined that they could not do without M. de Lesseps.

Reluctantly, then, for the beggarly sum of ten millions of francs, reduced to that amount, as they said, through his obstinacy, they knocked down their franchise to him, he, in fact, being the only bidder. The part least satisfactory remains to be told—the money has not been paid up. M. de Lesseps has, as it were, the *refusal* of this concession for two years on the basis of purchase. Now, some of the discontented and unreasonable shareholders repent of the sale and desire the return of the concession, having the idea that their interests have been sacrificed. It is almost wholly their affair; only one American, a citizen of New York, and a steady advocate of the San Blas route, appeared on the list as a participant in their joys and their sorrows, so far at least as ten thousand francs are concerned.

were concerned, the decision of the Congress did not seem to me likely to inflict injury other than a delay and an uncertainty as to the time of commencement of a great work.

In my Report to the Secretary of State, which many of you have doubtless read, I made the following deductions relating to the Congress:—

"That personal interests arising from a concession for the construction of a canal are unfavorable to a relative consideration of natural advantages as between two or more routes; that such personal interests did exist was quite apparent from first to last; and the 'concession' was frequently partially discussed or alluded to, especially in the committees or sub-committees.

"That the discussion in Paris has shown that hereafter, in the examination of the question, only the Nicaragua and Panama routes need critical examination, and that sufficient information exists as to all other routes.

"That the canal à niveau by the Isthmus of Panama, either with or without a tunnel, has been shown to be hopelessly impracticable, if considered as a commercial question.

"That a general and special knowledge now exists among European engineers relative to the subject of a ship canal across the American continent, which did not exist prior to the assemblage of the Congress in Paris.

"In view of actualities, it seems proper that the Government of the United States should consider the question of the inter-oceanic ship canal as still undetermined, notwithstanding the Report of its Commission on the subject, which has received acceptance by the people of the United States, and by our able civil engineers, inasmuch as it has not received a criticism.

"Should this be regarded as advisable, it would seem necessary to form a Commission of the ablest engineers of our Army, and to invite the ablest civil engineers of our country, and as well invite all the Governments who were represented at the Congress in Paris to send their engineers, all to join in full discussion, and having equal powers, with the view of removing it from all extraneous influences, of 'concessions,' or other objects than the consideration of the construction of a ship canal across this continent, capable of fulfilling the demands of the world's commerce, under the most economic conditions."

I have learned that the suggestion as to a Commission was maturely considered by our Government, and was regarded as unnecessary, in view of a supposed unanimity of the people of the United States in favor of the Nicaragua route.

It seems to me, however, that this fact, which I think undoubted, does not do away with the great advantage of the discussion of the subject by the ablest engineers, especially if, after a close study of the Nicaragua and Panama routes as presented by the surveys, they should visit both localities for the purpose of verifying any part of the work desired, and of the existence or non-existence of material for construction, and the methods which could best be employed in the execution of the work. This done, so far as human action can go, the question will be presented with the least possible condition of error, the locality where the canal should be made, or whether a canal should be made at all.

When it seemed to me that our Government was not disposed to call a Commission, I wrote, at the suggestion of a gentleman of position and influence, to Sir John Hawkshaw, presenting the advantage that the subject would derive from a personal inspection of the Nicaragua and Panama routes, either by him, or some able engineer appointed by the Society of Civil Engineers of Great Britain, accompanied by another appointed by the Society of Civil Engineers of France, and also another by our own Society, or, if our Government thought proper, to detail General Weitzel, U. S. Engineers, or any other competent officer who had large experience in hydraulic works. As yet I have received no reply.

The public is aware of the willingness of General Grant to assist in this great work, under such conditions of organization of a company, and of a concession, as would enable it to be prosecuted vigorously and effectively. He has given the subject his careful attention for years; is well satisfied as to the route, which possesses a certainty of realization by development; he appreciates fully the great importance of the construction of the ship canal for the commerce of the world, and especially for the full development of our west coast.

It is gratifying to observe that there is an universal expression of opinion as to the advantage which the construction of the canal would derive from having General Grant at its head. The expression is unanimous that it would ensure an economical, intelligent, and vigorous prosecution of the work, and its completion within the shortest time, and that it would have all of the conditions of practical utility and permanency that could be secured.

Recent information from the most reliable sources gives the assurance that the intelligent people and Government of Nicaragua are in entire accord with this movement, and, instead of embarrassing the question with impossible conditions, will do all in their power to forward the great work.

To sum up the whole matter, we may well desire that our countrymen should know what canal route will best serve the commerce of the world, in which our countrymen are so largely interested. This is eminently a question for the ablest engineers to pronounce upon; exact information will be presented in a prepared form by Civil Engineer Menocal, U. S. Navy, for discussion by the Society of Civil Engineers of the United States, showing the quantity of work that will be found necessary on the Panama route at the ocean-level, and also by way of Lake Nicaragua, with a lockage of 107 feet above the ocean-level. It is really not a question of what we may desire, but actually only of what nature, whose forces are ceaseless and tireless, will permit. To enter into an ill-advised struggle with them is to be defeated in the end, at whatever cost or continued effort. The labor and expense of constructing a ship canal under the most favorable conditions presented by nature will be great, but the result attained will be the grandest that man is capable of achieving for the amelioration of the commerce of the world.

Through this discussion we may hope that all of the advantages, as well as difficulties, positive and relative, on these routes, will be fairly developed, not upon fanciful presentations, but upon sufficient information through calculations.

I have taxed your patience in an endeavor to show the present aspect of the ship canal question, and have now only to point to the importance of the forthcoming calculations in detail, and the irrefragable results obtainable from their full and fair discussion.

[The greater portion of the foregoing paper was read on the evening of the 9th of December, 1879, before a large and very representative assembly of the American Geographical Society and its visitors by Colonel Theodorus Bailey Myers, of the Council of the Society. After expressing his regret that Rear-Admiral Ammen was unable to be present in person, and stating that his paper had been forwarded before the opening meeting, but was deferred to that of the Earl of Dunraven, who was in haste to return to England, Colonel Myers further prefaced the reading of the paper with the following remarks:—

"It seems proper to make an explanation on behalf of the Council of the position of the Society on the inter-oceanic canal question. Its hall and Journal have been open for years for its discussion. Foreseeing that all attainable information on the subject would soon be necessary, a Committee of the Council, consisting of Mr. Clarence King (Director of the United States Survey), Mr. Francis A. Stout, (Commissioner for the New York State Survey), and myself, by memorial and personal attendance at Washington, urged Congress, last winter, to print the surveys and statistics connected with the Government work on the Panama and Atrato-Napipi routes executed long before. This was only completed in a temporary form in time for the use of the Congress at Paris, and has recently been officially published. Access to it could not be had by the Society before that time, and was refused to individuals. In sending representatives to that Congress, the Society, therefore, claimed to take no part in the decision of an important question, of the merits of which, for these reasons, they could be but partially informed, but only as an appreciation of its importance, and to acquire information. Naturally they could not express an opinion without the time for study of prior details, nor could they expect that their representatives, during its brief and exciting session, should become able to do so. Two of these representatives have since given to the public their conflicting views on the plan presented by M. de Lesseps,—Dr. William E. Johnston, residing in Paris, by his able written report to our President, Chief-Justice Daly, received in the vacation, and printed in the Journal and through the press; and Mr. Nathan Appleton, in reading at about the same time a paper before the Board of Trade, supplemented by a communication to be read this evening.

"Rear-Admiral Ammen, as Chief of the Bureau of Navigation, having been charged with the fitting out of the American explorations, and as a member of a Commission formed by the Government for the consideration of its plans, having studied their results, has labored under no such difficulties. Those who know his capacity and devotion to any duty will believe his to be at least an educated opinion. Knowing that other conclusions will be advanced, he has authorized me to say that he is prepared to sustain it, and to reply to them if brought to his attention in the public press that he considers the subject worthy of exhaustive, if competent, discussion.

"On one so important, those present will, it is hoped, patiently submit to an extended discussion, caused by a desire to entertain all opinions, and open its merits to a free investigation."]

INSTRUCTIONS TO REAR-ADMIRAL DANIEL AMMEN AND CIVIL ENGINEER A. G. MENOCAL, U. S. N., DELEGATES ON THE PART OF THE UNITED STATES TO THE INTER-OCEANIC CANAL CONGRESS, HELD AT PARIS, MAY, 1879, AND REPORTS OF THE PROCEEDINGS OF THE CONGRESS.

MR. EVARTS TO REAR-ADMIRAL AMMEN.

DEPARTMENT OF STATE,
WASHINGTON, April 19, 1879.

REAR-ADMIRAL DANIEL AMMEN, U. S. N., WASHINGTON, D. C.:

SIR:—The President having appointed you to be a Commissioner on behalf of the United States to attend an International Conference, which is to assemble at Paris on the 15th of May proximo, under the auspices of the Geographical Society of Paris, for the purpose of considering the various projects of an inter-oceanic canal across the American Isthmus, I have the honor to acquaint you officially with the fact of such appointment. It is also incumbent upon me to give you certain instructions for your guidance in the execution of the President's wishes.

The importance and magnitude of the projected enterprise are such as to command earnest attention, especially on the part of those countries whose trade is to be affected in a marked degree by the success or failure of the scheme. This Government, in the interest of its rapidly growing commerce, not only between its own Atlantic and Pacific shores, but with the other American States on the western coast of the continent, deems it advisable to keep itself well informed on the subject, and also to give any useful information in relation thereto to other Governments interested in the scheme of inter-oceanic communication.

You are accordingly instructed to attend the Conference of the International Commission concerning the opening of an inter-oceanic canal through the American Isthmus, to be held at Paris next month, and you will be expected to carefully watch its progress and results, and report them to your Government. You will take part in the discussions of the Conference, and will communicate such scientific, geographical, mathematical, or other information as you may possess, and as is desired or deemed important. In this work you will be assisted by Civil Engineer Anecito G. Menocal, of the United States Navy, who has been detailed and appointed a Commissioner for the purpose, with like powers.

You will, however, have no official powers or diplomatic functions. You will hold no official communication with the officers of the French Government, except such as may, by virtue of their connection with the French Geographical Society, or as delegates *ad hoc*, take part in the proceedings of the Conference. You are not authorized to state what will be the decision of the Government of the United States in regard to the points involved, or the line of action it will pursue.

The Conference is understood to be not one of diplomatic representatives of the respective Governments, but rather a gathering of scientific men and public officers whose experience and research render it desirable that they should have an opportunity for the exchange of information and of views. Your well-known wide acquaintance with the subject proposed to be discussed makes it peculiarly fitting that you should be selected to meet other distinguished engineers and officers who have given like attention to the matter.

You are furnished herewith with such documents and records as the files of this and other departments contain that may be of interest and importance. .Your own long familiarity with the subject and careful study of it in all its bearings render it unnecessary to give you further instructions on this point.

I am, sir, your obedient servant,

WM. M. EVARTS.

REAR-ADMIRAL AMMEN TO MR. EVARTS.

WASHINGTON, D. C., June 21, 1879.

SIR:—I have the honor to inform you that Civil Engineer A. G. Menocal and myself reached Paris on the morning of the 14th of May, and lost no time in paying our respects to M. Ferdinand de Lesseps, upon whose invitation, as President of the Geographical Societies of Paris, the Department was pleased to direct our attendance at the convocation on May 15th.

We were received by M. de Lesseps with great courtesy, who spoke of the very general attendance of different nationalities, and his gratification thereat. A general and agreeable conversation occurred relating to the inter-oceanic canal question.

On the afternoon of the 15th the convocation was called to order, in the building of the Geographical Society, by Vice-Admiral Ronciere le Noury. After a few remarks, he turned over to M. de Lesseps the office of presiding, who made some remarks, and was followed by M. Bionne, the Secretary of the Society, who read a paper on "The State of the Inter-Oceanic Canal Question."

The organization was then completed, five vice-presidents appointed, I being named first and seated on the right of M. de Lesseps.

After a call of the members, an assignment to committees was made, all of which will appear in full on the journal of the proceedings when published.

The assemblage was then adjourned to meet as committees at 9 A.M. on the following day (16th), no general session to occur until Monday, May 19th.

The technical question on the 16th was discussed in the large hall in which the assemblage had been organized the previous day. M. Doubrée, the Chairman of the Committee, called upon me to produce the maps and plans brought by us; owing to the weight of the package it had been sent by express, and did not arrive in time for presentation until the following morning. In the meantime Commander Selfridge, U. S. N., had the attention of the Committee, and presented his "plan and estimates," based upon the surveys of Lieutenant Frederick Collins, U. S. N., over a region known as the Atrato-Napipi route. A preliminary instrumental examination of the bed of the Napipi River had been made prior to this survey by Commander Selfridge, upon which he had based a supposititious location, and made "estimates" without ever having been over the ground.

At two o'clock, when the Committee again met, I stated that there was no one to whom we were more indebted than to the able and energetic officer who had just concluded, for the examination of lines extending over a wide region, and requiring great labor and privation to execute. So far, however, as the route was concerned, I had to say that the survey and the discussion of that route had been made by Lieutenant Collins himself, who was not the inferior to any officer of our Navy. He had been placed in communication with the ablest civil engineers in our country, from whom he had advice and assistance; the official maps, plans, estimates, and reports made by him formed a part of the information which it was my duty to lay before the Congress. I was sorry to add that they were not as favorable to the execution of the work, and as to cost, as that which had been presented by Commander Selfridge. He replied, stating that Lieutenant Collins had been his subordinate for three seasons, and had been again sent to that region at his (Selfridge's) request; that he had supposed, in the development of the plans, he would be consulted; as he was not, he had requested the data, and had made his plans. I could have replied very properly and satisfactorily, but did not out of respect to the time and character of the persons composing the Committee.

On the following morning, May 17th, I read my paper, which had been submitted for your revisal. Mr. Menocal then very clearly and ably presented the survey of the Nicaragua route, and explained several of the methods by which he proposed to overcome difficulties in the execution of the work, particularly as to the improvement of the harbor of Greytown, and as to the foundation and construction of the different dams; and finally, in answer to whatever difficulties were suggested. There was evident surprise on the part of very able and competent engineers at the cleverness in design and completeness of detail for execution of what had been presented, and he received warm congratulations. This presentation occupied several hours.

The afternoon was taken up by Lieutenant L. N. B. Wyse, of the French Navy, who placed on the stand for explanation his development for a canal in the vicinity of *the line of the Panama Railroad*. I was not surprised at this change of base from the Tuyra-Tupisa route, having read his last Report, which was given me two days before. His discourse was general, and referred to all of the proposed lines, and occasionally something about the Isthmus of Panama.

M. Felix Belly, whose name will be remembered in connection with the Nicaragua route in 1858, then claimed the attention of the Committee.

On Monday, May 19th, a general session was held, and several reports of the committees read, copies of which will be procured when possible and sent to the State Department.

Sir John Hawkshaw arrived, much to my gratification; his authority as an engineer is of the highest order. With men present of his reputation, character, and ability, the discussion of the question will be of great prospective value, whatever the Congress may fail to decide, or, rather, determine.

The afternoon was taken up by alternate explanations of Lieutenants Wyse and Reclus, of the French Navy, in their development of a ship canal in the vicinity and along the general line of the Panama Railroad *à niveau*, that is to say, on the ocean-level, with and without a tunnel, and as well in stating objections at any time to the Nicaragua route. They were prolix, and their data were not at all sufficient, being, in fact, mainly on the railroad line levels, and a few cross sections run by Lieutenant Reclus. From the fact that Lieutenant Wyse had quite abandoned even the discussion of the Tuyra-Tupisa route and taken up the line of the Panama Railroad, the cause of his anxiety in February last, when in the United States, to obtain our surveys of that canal route became quite apparent.

On Tuesday, May 20th, a general session was held. Commander Selfridge had the floor again to finish, as he said, his explanation of the Atrato-Napipi route, which continued for two hours.

When he concluded, I stated that I had the very carefully prepared development of Lieutenant Collins's surveys, which was an actual location made by that very able officer, and was over the route which Commander Selfridge discussed; that Collins's plans and estimates had been made after consultation with able engineers, and that, when the opportunity offered, I would present them in detail. They, however, did not present the favorable features for canalization which were assumed to exist by Commander Selfridge.

Mr. Menocal was then invited to explain the surveys and plans of the Panama route, which he did with great clearness, evidently much to the satisfaction of Sir John Hawkshaw, so far, at least, as a comprehension of the points involved was concerned. He exposed the hopelessness of an attempt to make a ship canal on that route *à niveau;* pointed out, beyond controversy, that if so made there would be a cataract of the River Chagres at Matachin of 42 feet, which in periods of floods would be 78 feet high, of a body of water that would be 36 feet deep, with a width of 1500 feet.

The surprise and painful emotion on the part of those who had plans *à niveau*, and of their very many friends in attendance, can hardly be conceived. The fact stared them in the face that the plans which they had presented so confidently for adoption were absolutely impracticable. There was, however, after a day or so, a presentation of " plans " and estimates of the cost of execution, quite independent of a sufficient knowledge of the topography upon which only could they be properly based.

Mr. Menocal went on to explain how a water supply was obtainable, and that, owing to the floods of the Chagres River, it was impossible to lower the bottom of the canal below the height of the aqueduct, as proposed by him, crossing the river at Matachin.

When he concluded his presentation of the Panama route, he stated that he would be happy to answer any questions which might be proposed. Not a question was asked, although when he commenced he was so interrupted that Sir John Hawkshaw suggested that he should be allowed to proceed without interruption, and make explanations afterward. An adjournment occurred soon after until 2 P. M.

I learned from Mr. Menocal that at the meeting of the Committee in the afternoon great confusion and violent actions preceded the appointment of two sub-committees, the one to report on tunnels, such as proposed by Lieutenant Wyse on the Panama route, the other to discuss the question of the practicability of canal locks, as proposed on the Nicaragua route. Preceding and during the formation of the committees, there was, on the one hand, a strong demand that Mr. Menocal should be on the Committee of Locks, and, on the other, a violent demand that he should not be. Very properly, in my opinion, he requested that he should not be appointed. After the committees were made, however, the one referred to demanded of the Chairman the attendance of Mr. Menocal, which was formally granted.

On the 21st a general session occurred, and the Report of the Committee on Tunneling was read; this Committee was composed of the ablest engineers in Europe in that branch of engineering. A copy of that Report will at the earliest date be obtained and sent to the Department.

It is sufficient to state in brief that it arrived at the impracticability, at whatever cost, of constructing a tunnel *à niveau*, that is to say, to secure navigable waters at the ocean-level.

Sir John Hawkshaw then proceeded with brevity and great force to give his views on the construction of a ship canal at the ocean-level, as proposed by Lieutenant Wyse, along the general line of the Panama Railroad. A copy of his remarks is appended, marked A. I may properly add that the report of the sub-committee on tunneling before referred to, as well as the deductions of Sir John Hawkshaw, support fully the ideas advanced in my paper, read before the American Geographical Society of New York in November last, as to a canal and tunnel *à niveau*.

In commenting upon the proposed tunnel, Sir John remarked that considered *as a culvert*, and taking the volume of water as given by Lieutenant Reclus in periods of floods, which was much less than

SHIP CANAL QUESTION. 75

given by Mr. Menocal, it would require at least another tunnel of the same dimensions *to serve as culverts* to pass the water.

Lieutenant Reclus then began an argument in a desultory way, and when the Chairman requested that he would confine himself to the subject under discussion, Lieutenant Wyse in a very excited manner said their plans were constantly attacked, and they were not permitted to defend them.

Commander Selfridge then said that the remarks made by Sir John Hawkshaw in relation to the Chagres River were not applicable to the Atrato-Napipi route, and endeavored to enter into a further discussion of its merits. I stated that I would suggest the advantage of discussing the carefully prepared plans of Lieutenant Collins along the lines of actual location, which were the best that could be found in months of labor, instead of lines drawn at will by Commander Selfridge, involving uncertainty of execution and an entire absence of elements of calculation, as every engineer would recognize. The meeting then adjourned.

On the 22d the sub-committees were in session, and Mr. Menocal was in attendance to give such information as was required.

At 9 A. M. of the 23d a general session occurred. After the usual preliminary proceedings the report of the Committee on Navigation was read. A somewhat lengthy address was then made by the President, M. de Lesseps, partly in reply or in relation to some preceding remarks of other speakers, and partly in relation independently to the inter-oceanic canal question, the full import of which can only be known when it can be carefully read.

This was replied to with some warmth by Mr. Peralta, the minister of Costa Rica to the United States, after which Commander Selfridge arose and stated that as an American citizen he protested against the supposition being entertained that the people of the United States had any preference between three or four inter-oceanic canal routes ; that they would accept truly and loyally whatever decision was arrived at by this Congress, in whose wisdom they would have full faith. I made no reply, inasmuch as it would have been difficult to explain, as well as unimportant, how far his assertion was correct or the reverse.

On the morning of the 24th, as the sub-committees were at work, I thought it worth while to say formally to their chairmen that I had no objection to the Atrato-Napipi route being presented by Commander Selfridge, provided its consideration was made upon the actual lines of location made by Lieutenant Collins, who had spent months in making them, and that they were undoubtedly the best that the nature of the country permitted. Seeing Commander Selfridge in the presence of Mr. Jackson, one of the secretaries of the Congress, I asked his attention whilst I stated the preceding facts for transmission through the secretary to the sub-committees.

Commander Selfridge stated that he was a member of this Congress, invited by M. de Lesseps; that when before the committees he had pointed out " where he diverged from Lieutenant Collins's lines to get a little nearer the river and thus diminish the cutting, and that he thought his opinion better than mine, as I had never been in that region." I replied that my opinion was based upon the careful instrumental location of lines made by Lieutenant Collins, and his opinion was based upon drawing lines at will for discussion.

Owing to proposed modifications by the technical sub-committees of the cross-sections, locks, and other conditions of the Nicaragua Canal, Mr. Menocal was kept very closely employed in making them, and was most efficiently aided by Lieutenant-Commander Gorringe, United States Navy, who opportunely was passing through Paris, under orders for home. At my request he kindly tarried as long as necessary.

Sunday, May 25, was busily employed by the sub-committees in calculations to meet requirements of construction other than those officially presented by us.

On Monday, May 26, the technical sub-committees reported to the full Committee. There was, however, a very general attendance of delegates who did not belong to the Committee, and a very great deal of interest and feeling manifested, amounting, at times, to disorder, when the reports were read. I visited the room both in the morning and afternoon session, but preferred to be generally absent.

On the morning of the 27th a general sitting occurred, and the statistical and other reports were read, after which an adjournment took place, the Technical Committee to meet at 2 P. M., which meeting I did not attend. I was informed, however, that the able engineers were very generally in favor of the Nicaragua route, and that nearly, if not all, of the French delegates, other than the engineers, were in favor of the Panama route. As it is eminently a question for engineers to settle, the votes of other persons would seem to me to have significance rather as to to personal interest than to relative practicability of routes considered on their abstract merits.

From the first sitting it was quite apparent that there were two parties of what we would call "speculators," the one represented by Mr. Blanchet, who had an unconfirmed grant from the Nicaraguan Government, and Lieutenant Wyse, of the French Navy, who had a grant from the Colombian Government, embracing, with a reservation, the right to construct a ship canal over any part of her territory, the reservation applying, as I understand it, to the already conceded right of the Panama Railroad. The presumed grant to Lieutenant Wyse was published in the New York *Herald* nearly one year ago. Lieutenant Wyse has the powerful support of M. de Lesseps. I need hardly add that through the geographical societies of Paris, and the method of appointing "delegates" to the Congress, the latter is quite able to have any desired majority on a vote relating to the respective merits of the Nicaragua and Panama routes.

The advocates of the Nicaragua route were disposed to regard Mr. Menocal and myself as an accession to their ranks, a position that we have persistently refused to accept, recognizing the fact that the mere preference of opinion in relation to the superiority of the Nicaragua route did not make it a duty to become advocates except by inference, and the presentation of facts which would support that opinion. The absence of exact information, and perhaps the prejudices of the ablest as well as of the engineers in general present in the beginning of the discussion, at least made them the tacit supporters of the ideas of M. de Lesseps as to a canal *à niveau*. It was quite apparent, as the routes were presented and discussed, that the able engineers generally ranged themselves on the side of the Nicaragua route. The very able presentation of both routes by Mr. Menocal, and the opinion expressed by Sir John Hawkshaw, quite disturbed the equilibrium of the Panama route advocates, as will appear when the reports of the sub-committees are published and come to hand.

On the 28th of May the technical sub-committee met to discuss a new "plan" based upon making a high dam, higher in fact than the surface of Lake Nicaragua above the ocean-level, at some point across the Chagres River, with the intention of flooding a considerable tract of land in forming a large lake, and thus ameliorating the destructive effects of floods. Ten days before, this idea had not entered the mind of man; it was in effect the resultant of the exposition that the canal *à niveau* was hopelessly impossible without amelioration and that no other solution exists along that route except the adoption of the plan of canal presented by Mr. Menocal at a height sufficient to allow the Chagres to discharge its floods beneath the aqueduct. Of course, the serious consideration of such a work as forming this large artificial lake could only be made properly, after a very thorough examination of the topography at points most favorable for natural abutments, as to height and length of dam required, and above all as to foundation. In presenting the case, however, to spare it from ridicule, it was necessary to assume that a canal *à niveau* is the demand of the commerce of the world at any cost. Whatever possibilities a special and close survey may develop, it is quite certain that the natural conditions are now unknown upon which the predications of these " plans" are based.

In the discussion of a sub-committee, M. Gauthiot stated that the tolls of a vessel of 4500 tons through the Suez Canal would be 45,000 francs, but through the proposed Panama Canal they would be 120,-000 francs. Not having further information on the subject, I presume this to be the fact, the more as Lieutenant Wyse, who secured the concession, spoke of the faculty of charging 20 or 30 francs per ton, which, if established, would virtually exclude a ship laden with grain bound from San Francisco to Europe.

It seems apparent to me that the construction of a ship canal should be regarded simply as a commercial question; that very high tolls would, in a great degree, take away from its usefulness, and were a canal constructed across the American continent by persons who had dominant interests in the Suez Canal, the American Canal would be subordinated to the interests of the Suez Canal. Thus, with a difference of tolls as above given, all vessels from Northern China and Japan bound to Europe would pass through the Suez Canal, and all vessels bound for our Atlantic coasts would pass through the canal across this continent at rates which would nullify in a great degree the proposed commercial benefit.

On the morning of the 29th M. de Lesseps held a meeting of the vice-presidents and other persons to determine the manner of voting and to submit the reading of the Report of the Technical Committee, as also a resolution. There was a long discussion, and it was at length agreed upon that in addition to a single vote on the resolution to be submitted to the Congress of yes or no, or of abstaining from voting, any delegate should be permitted within three days to give, in writing, for record, the reasons which governed his action.

At 1.30 the final full meeting of the Congress took place; the Report, *résumé*, and resolution were read, and the yeas and nays taken on the latter, resulting in a vote of abstention of 98 members, out of 135 as given in the list—75 voting yes, 8 no, and 16 abstaining. The character of the voters and those who absented themselves will appear in the Report of Civil Engineer Menocal. I abstained from voting on the ground that " only able engineers can form an opinion, after careful study, of what is actually possible, and what is relatively economical, in the construction of a ship canal."

The text of the resolution is as follows:—

Le Congrés estime que le percement d'un canal interocéanique à niveau constant, si désirable dans l'intérét du commerce et de la navigation est possible, et que le canal maritime pour repondre aux facilités indispensables d'acces et d'utilization qui doit offrir avant tout un passage de ce genre devra être dirige du Golfe de Simon à la baie de Panama.

The hall was densely crowded, many ladies being present; about one hundred members or delegates and three to four hundred other persons: whenever a vote of "yes" was given, especially by some one who had more or less opposed the conclusion, a very enthusiastic clapping of hands occurred which would hardly have been the case had the audience regarded the selection as depending wholly on natural conditions or advantages, or on physical causes. The Congress then adjourned.

It is proper to acknowledge the courtesy and consideration shown to myself and to Mr. Menocal by M. de Lesseps and by many of the delegates, many of whom were gentlemen of eminence, as M. Ceresole, late President of the Helvetic Confederacy.

The conclusions deducible from the above I regard as follows:—

That personal interests arising from a concession for the construction of a canal are unfavorable to a relative consideration of natural advantages as between two or more routes; that such personal interests did exist was quite apparent from first to last; and the "concession" was frequently partially discussed or alluded to, especially in the committees or sub-committees.

That the discussion in Paris has shown that hereafter in the examination of the question only the Nicaragua and Panama routes need critical examination, and that sufficient information exists as to all other routes.

That the canal *à niveau* by the Isthmus of Panama, either with or without a tunnel, has been shown to be hopelessly impracticable, if considered as a commercial question.

That a general and special knowledge now exists among European engineers relative to the subject of a ship canal across the American continent, which did not exist prior to the assemblage of the Congress in Paris.

In view of actualities, it seems proper that the Government of the United States should consider the question of the inter-oceanic ship canal as still undetermined notwithstanding the Report of its Commission on the subject; which has received acceptance by the people of the United States, and by our able civil engineers, inasmuch as it has not received a criticism.

Should this be regarded as advisable, it would seem necessary to form a Commission of the ablest engineers of our Army, and to invite the ablest civil engineers of our country, and as well invite all the governments who were represented at the Congress in Paris to send their engineers, all to join in full discussion, and having equal powers, with the view of removing it from all extraneous influences, of "concessions," or other objects than the consideration of the construction of a ship canal across this continent, capable of fulfilling the demands of the world's commerce, under the most economic conditions.

Very respectfully, your obedient servant,

DAN'L AMMEN,
Rear-Admiral United States Navy.

Hon. WM. M. EVARTS,
Secretary of State.

REMARKS OF SIR JOHN HAWKSHAW AT THE INTER-OCEANIC CANAL CONGRESS IN PARIS, MAY, 1879.

(A.)

With regard to the question whether the canal should be constructed with or without locks, the following points occur to me :—

If the canal is to be without locks its normal surface-level would be that of the sea, and its bottom-level, say eight metres lower. This being the case, the canal would receive and must provide for the whole drainage of the district it traversed.

Therefore it would be necessary to ascertain the volume of water that would drain into the canal before it would be possible even to determine the sectional area of the canal.

If the canal have a less surface-fall than the river, as it would have, it must have a larger sectional area to discharge the same volume of water.

The average section of the river in a flood at Mamei was ascertained by Mr. Reclus (page 175) to be 1310 square metres. This would require a canal, if it were eight metres deep, to be 160 metres wide.

The waters of the Chagres would have a tendency to flow toward the Pacific, that is, through the tunnel, as the distance is less and the fall greater than to the Atlantic.

It seems to me that the dimensions of the tunnel, if it has to serve for both the river and canal, would be too small. Mr. Menocal's estimate of the volume of the Chagres in time of flood would much more than fill the tunnel; and in any case the whole section of the tunnel is only half that of the river in time of flood, as given by Mr. Reclus.

During the construction of a canal at the sea-level difficulties would arise in providing for the drainage, which would affect both time of execution and cost to an extent that could hardly be ascertained in advance.

If, from such considerations as the foregoing, it should be concluded that the canal should be so constructed as to retain the rivers for natural drainage, then recourse will have to be had to locks.

In that event there can be no difficulty, in my opinion, in carrying on the traffic with locks properly constructed, provided there is an ample water supply, which would be a *sine qua non*.

SHIP CANAL QUESTION. 81

MR. MENOCAL TO MR. EVARTS.

UNITED STATES NAVY YARD, WASHINGTON,
CIVIL ENGINEER'S OFFICE, June 21, 1879.

SIR:—In compliance with instructions from the Department of State, dated April 19, 1879, I have the honor to respectfully submit herewith a Report of the proceedings of the Technical Committee constituted by the International Congress convened at Paris, France, for the purpose of discussing the question of an inter-oceanic ship canal across the American Isthmus, and to which Committee I was assigned; embracing, in substance, the views of foreign engineers and others; together with such statements as I felt called upon to make from time to time during the various discussions and deliberations of the Congress relative to the project.

All of which is most respectfully submitted.

I am, etc.,
A. G. MENOCAL,
Civil Engineer, U. S. N.

UNITED STATES NAVY YARD, WASHINGTON,
CIVIL ENGINEER'S OFFICE, June 21, 1879.

SIR:—Rear-Admiral Daniel Ammen, U. S. N., has been good enough to show me his Report to the Department of State of the proceedings of the International Congress, convened at Paris on the 15th of May last, to discuss the question of an inter-oceanic ship canal across the American Isthmus. As one of the delegates appointed by the President of the United States to that Convention, I was instructed by the Department of State to report from time to time the transactions of the Congress, but the statements of the Admiral seem to me so clear and complete a presentation of the facts that it would be superfluous to submit a full Report. I beg leave, therefore, to confine myself to a statement of the workings of the Technical Committee, of which I was a member, and to which all projects for an inter-oceanic canal were presented for discussion.

The first meeting of the Committee took place on the morning of the 16th of May. Admiral Ronciere le Noury proposed Mr. Doubree for President, Messrs. Dirks and Kleits for Vice-Presidents, and Messrs. W. Kuber de Maere, Linmander, and Emilie Muller as Secretaries, and they were so elected by acclamation.

On the organization of the Committee, the President proposed to commence the examination of the different projects, beginning with those of the Government of the United States. Rear-Admiral Ammen was, accordingly, requested to submit such plans as were under his charge, and in his absence I thought proper to state that the trunk and case containing the Reports and maps of the different surveys conducted by the Government of the United States had been unavoidably delayed on their way from Liverpool to Paris, but that they were expected to arrive on the evening of that day, and that I hoped they would be ready for presentation on the morning of the following day.

The President then requested Commander T. O. Selfridge, U. S. N., to present to the Committee the projects of which he had made a special study.

Commander Selfridge proceeded to explain, at length, his modified plans and estimates for a ship canal by the Atrato and Napipi rivers, based, as he stated, on the last surveys made by Lieutenant F. Collins, U. S. N., and his notes from previous surveys conducted by himself. The length of the work as proposed by him, as well as that of the tunnel, seems to be the same recommended by Lieutenant Collins, according to the line actually located by this able officer; but Commander Selfridge claims that by locating the line in closer proximity to the rivers Napipi and Doguado a lower profile can be obtained, and the estimated cost of construction reduced from $98,196,894, as given by Lieutenant Collins, to $53,000,000. Commander Selfridge spoke at length on the practicability of effectually and permanently removing the bar at the mouth of the river Atrato by adopting the same system of jetties now in course of construction at the mouth of the Mississippi River by Captain Eads, and added that, with the aid of fifty men, he could in three months obtain 15 feet depth of water on the bar at the Atratro, now 2500 feet in length and less than 3 feet in depth of water. He also spoke of the necessity of providing any canal built on the Isthmus of America with ample facilities for surface-drainage, and believed that keeping the surface of the water in the canal above that of the ground was of great advantage to avoid injury from floods in the rainy season. He was then asked by Mr. Laroche what provision he had made to that effect in the project for a canal under consideration, to which he replied that he had not directed his studies to that special point, which particularly belonged to the competency of the engineer.

Commander Selfridge stated at the close of his remarks that he was engaged in the preparation of plans and estimates for a canal without

SHIP CANAL QUESTION. 83

locks by the same route, which he expected to submit to the Congress shortly.

Mr. Blanchet was then allowed to present his plans for a canal by Lake Nicaragua, on which he dwelt at length. His project consists in a modification of that contained in the Report of Commander E. P. Lull, U. S. N., to the Navy Department of the operations of the United States Surveying Expedition to that country in 1872–1873. Mr. Blanchet has made two trips to Nicaragua, in the interests of an association which he represents, with the view of obtaining a concession from that Government to build the canal. He has not, however, made any surveys; and his project is based either on the information obtained by the United States surveys, or on mere speculations as to the topography of the ground. Little weight can be attached, therefore, to his statements as to natural conditions; but even should they be verified by proper examination of the country, his plan would be involved in much uncertainty as to practicability of execution, cost, and permanency of the works proposed.

He was not prepared to present plans for the construction of harbors at Brito and Greytown, the two ends of the proposed canal; and when requested by the Committee to submit estimates for these works, he referred to a Report made by me to the Government of Nicaragua in 1876, which he accepted as the best solution of the problem. To this extent Mr. Blanchet seems to give credit to the labors of others, which he so readily appropriates. His estimate of cost of construction is $38,146,352.

On the morning of the 17th Rear-Admiral Ammen made an exposition of the different projects for a ship canal, surveyed by the several expeditions sent out by the Government of the United States to Tehuantepec, Nicaragua, Panama, and Darien, about to be presented to the Committee for consideration.

I was then called upon to make a technical exposition of those projects. Referring to the proposed canal by the Isthmus of Tehuantepec, I stated, after consultation with Rear-Admiral Ammen, that it was regarded of so little merit, as compared to other routes more to the south, that I deemed it unnecessary to go into a full explanation of the lines; that I would submit for the information of the Congress a number of Reports of Commodore Shufeldt, under whose command the survey had been conducted, and that, should any of the delegates desire any information as to any particular point, I would be pleased to convey it to the extent of my ability. I proceeded then to present my views relative to the proposed canal across Nicaragua, as located by the United States Surveying Expeditions of 1872-1873, under the command of Commander E. P. Lull, U. S. N., to which I was attached as chief engineer. The work was therefore quite familiar to me; a fact which, together with a knowledge of the country and its resources of all kinds obtained from actual observation during several years, made my task comparatively easy.

The Nicaraguan Canal has been generally considered by American engineers, since the results of the surveys under consideration were published, as the best and most practicable route. It involves less engineering difficulties than any other, and its estimated cost, based on an actual and careful location of the line, is so small, as compared to other routes, that, in a commercial sense, it seems to take precedence above all others.

I endeavored to give a clear and full description of the whole line, stating the area, water-shed, and elevation of Lake Nicaragua above the level of the sea, and the fact of its being the summit-level of the proposed canal; the name and location of the different trial-lines run from the lake to the Pacific, with a full description of those passing through the lowest depressions of the divide, and particularly of the one selected as the best, with the reasons therefor. I also described at length the work required on the lake to make it navigable for large vessels along the whole distance across to its outlet; the River San Juan, with all its features in relation to the canal and the proposed works to make it navigable for a distance of 63 miles from the lake. I also described the nature of the country from the point where the canal leaves the channel of the River San Juan below its confluence with the San Carlos to Greytown; giving distances, number, and lift of the locks required on both sides of the summit, together with such works of detail as are essential to the successful construction and permanence of the canal.

The want of good harbors at both ends of the proposed canal across Nicaragua had been mentioned by many as the weakest feature of that project, and to that particular point I had devoted much study during the last four years. There was a marked desire on the part of many of the delegates to hear what I had to say in reference to those parts, and especially that of Greytown. I explained with much care the plans proposed for the construction of a harbor at Brito sufficiently large to meet the requirements of the canal, and answered several questions asked by different members as to natural conditions and facilities for the execution of the works. I then proceeded to describe the present condition of the harbor of Greytown;

the changes that had taken place during the last forty years I illustrated by maps prepared for that purpose, pointing out the causes that had produced those changes, and brought about the closing of the harbor. During this exposition I invited criticism. As I was professionally interested in the solution of this important problem, I readily answered all questions relating to the subject, and finished my remarks with an explanation of the plan proposed for the restoration of the harbor, with provisions for its permanency. I have the pleasure to say, in this connection, that these plans were shortly after referred to two sub-committees, composed of thirteen of the most eminent engineers present, and that they were approved in all particulars.

I was called upon to explain the system by which the proposed dams across the River San Juan were intended to be built, when I had occasion to submit the detailed plans prepared for said works, and to describe the manner in which I would proceed to overcome the difficulties connected therewith. The explanations seemed to have been received with general satisfaction. After referring to other works of detail, and to the abundance of building·materials, easy means of communication, etc., the presentation of that route was considered ended.

Lieutenant L. N. B. Wyse was then requested to bring before the Congress his project for a ship canal. He spoke in general terms of the inter-oceanic canal question; and after mentioning the different lines surveyed by the United States, he referred to the surveys and reconnoissances conducted by himself on account of the International Commission of Paris. Five different projects had been proposed between the Nicaragua and Atrato-Napipi routes, and of these he had had occasion to examine four. The one he failed to examine was that connecting the Bay of Aspinwall with that of Panama. After enumerating the advantages and disadvantages of the several routes studied by him, he arrived at the conclusion that they were more or less impracticable, and that a canal at the level of the sea by the Isthmus of Panama (the route he did not survey), with or without tunnel, meets all the requirements for a ship canal, and could be executed at a moderate cost. He then described the line of the proposed canal, stating that his information had been obtained from the plans of Mr. Garalla, made in 1843, for a canal with locks and tunnel (in a different direction from the line proposed by Lieutenant Wyse), and from the plans of the Panama Railroad. A canal at the level of the sea, such as proposed by Lieutenants Wyse and Reclus, has necessarily to be the ultimate drain or recipient of the extensive basins intercepted by the canal, including those of the rivers Chagres, Obispo, Grande, Trinidad, Trijolis, Gutun, and others. No provision had been made, however, to dispose of the surface-drainage, nor to ameliorate the effects in the canal by the rise and fall of the tide in the Bay of Panama, the latter amounting to about 24 feet.

I will have occasion to refer again to the project on the presentation of a canal with locks on this line.

On the 20th Commander Selfridge rose to state that since his arrival he had been able to ascertain that a large number of engineers were opposed to a canal with locks; that the more he considered the subject, the more he agreed with the French engineers, and that he had accordingly modified his plans of a canal by the Atrato-Napipi, so as to reduce the number of locks to two. After a few remarks by Rear-Admiral Ammen, I was called to explain the proposed canal by the Isthmus of Panama, as located by the United States Surveying Expedition of 1875. I described it with the same care and minuteness of detail as that of Nicaragua. I thought proper in this connection to state that the first investigations of the commander of the expedition and myself, on our arrival at the isthmus, were directed to ascertain the practicability of a canal at the level of the sea; that we found the elevation of the River Chagres at Matachin, in its normal condition, to be 42 feet above the sea-level, and in times of flood as high as 78 feet, with a width of channel of about 1500 feet. The canal would necessarily have to be located along the valley of this river, and as there was not (in this narrow valley) room enough for both the canal and the river, the latter had necessarily to be received into the canal under a fall varying from 42 to 78 feet, according to the magnitude of the floods. Add to this immense cataract the floods of the other rivers, tributaries of the Chagres below Matachin (some of them of large water-shed and considerable volume), and it would be easy to conceive that a canal at the level of the sea, as proposed, by Panama, would frequently be converted into an extraordinary torrent, impracticable for navigation. The cost of a canal under these conditions cannot be estimated, but it would be so great as to make it commercially impracticable. It was decided, therefore, to accept the locks as the only possible solution of the problem.

I was interrupted in my explanations by Lieutenants Wyse and Reclus, who wished to know whether our information as to elevation and rise of the Chagres had been obtained by actual surveys or from the natives; to which I positively replied that it had been obtained by careful lines of levels run from the Atlantic and Pacific oceans to Matachin, and along the bed of the River Chagres to sixteen miles above that place. On the afternoon of the same day, Mr. Garay, delegate of the Mexican Government, made some remarks advocating a canal by the Isthmus of Tehuantepec. Mr. Rebourt, engineer of the Saint Gothard Tunnel, followed with a discussion on the probable cost of a canal without locks by Panama, and showed by figures and diagrams that the cost of that work at such prices of labor as are paid in France, and supposing the work to be done out of water, would be 930,000,000 francs, and time required for its execution no less than nine years. This statement caused a long discussion, in which many of the delegates took part. It was at last proposed and agreed to by acclamation that two sub-committees be appointed: The first, on excavation in tunnel and open cut, dredging, and estimates; the second, on locks and cross-section for the canal. These sub-committees were instructed to report as soon as practicable, and to prepare comparative estimates of cost of all the projects thus far presented to the Technical Committee.

On the morning of the 21st Messrs. Cotard, Ruelle, and Lavalley, members of the first sub-committee, made some statements to the effect that the practicability of a canal at the level of the sea, as proposed, was involved in much uncertainty, and that the difficulties to be overcome were of such a character and of such magnitude that it would be impossible for any engineer to arrive at an approximate estimate of its probable cost.

Sir John Hawkshaw followed with a few remarks to the effect that the volume of the Chagres, in times of flood, would be more than sufficient to fill the entire cross-section of the tunnel (as proposed by Lieutenants Wyse and Reclus), which is only one-half that of the river, as given by Mr. Reclus.

That the necessity of keeping the excavation free from water would increase the cost and time of the execution of the work to a point that cannot be estimated beforehand.

That the velocity of the current in the proposed canal due to the tide alone would be about five miles an hour. "That, if from such considerations as the foregoing it should be concluded that the canal should be so constructed as to retain the rivers for natural drainage, then recourse must be had to locks. In that event there can be no difficulty, in my opinion, in carrying on the traffic with locks properly constructed, provided there is an ample water supply, which would be a *sine qua non.*"

These remarks from so eminent an engineer as Sir John Hawkshaw had much weight in the Committee.

Lieutenants Wyse and Reclus had, since the day before, commenced to suggest some means by which to overcome the difficulty of receiving the River Chagres into the canal in the form of a cataract. The remarks of Sir John Hawkshaw, however, made it only too apparent to them, that unless they succeeded in disposing of this difficulty in a manner acceptable at least to their friends, their project had to be abandoned. These gentlemen are *not* engineers, and however competent they may be in their profession, they lack that theoretical and practical knowledge without which a work of such magnitude as the one under consideration cannot be designed or executed.

On the 22d instant I was called by both sub-committees to give some additional information relative to the Nicaragua Canal. The sub-committee on locks requested me to submit plans and estimates for a system of three locks of 150 metres in length, between gates 20 metres wide and no more than 4 metres lift. Also, estimates for an increase in the width of the canal to 100 metres at the bottom, in curves and bends of the river of less than 3000 metres radius. It was apparently the intention of this Committee to so increase the cost of the canal with locks, *via* Nicaragua, as to make it less practicable in a commercial sense than the canal at the level of the sea *via* Panama, which had very strong advocates in that body, particularly its President.

I prepared in the evening of that day plans for a system of double locks of the dimensions given, which I submitted, together with the estimates required, on the following morning. I refused to make plans for the three locks on the ground that I considered it unnecessary, since a canal with single locks would allow the passage of no less than 24 ships a day, and one with double, more than double that number. According to my calculation, a ship could pass a lock (as I had designed them) in about one-half hour, but I have Sir John Hawkshaw's authority for stating that 20 minutes would be ample time for the operation. Supposing that 5,000,000 tons would pass through the canal during the year, which seems to be a liberal allowance, and that the ships will carry 2000 tons each, the average daily passage would be 7 ships. I believe, therefore, that in order to build the canal as economically as possible, so as to make it a financial success, single locks should be built at the start, and if at any time the traffic should increase so as to require additional facilities, a second series of locks can be built alongside the first ones, without any inconvenience or great cost.

On the 23d instant I was furnished by the first sub-committee with a drawing for a new cross-section for the canal proposed by the sub-committee on locks, and requested to compute the cubical contents of excavation resulting from this change and from an increase to 100 metres of the whole channel in both the river and lake. The new cross-section proposed I regarded as utterly impracticable, and seemed to have been designed for the only purpose of reducing relatively as much as possible the cubical contents of excavation in the canal without locks by Panama. The proposed slopes in cuttings of nearly 300 feet in depth is one-tenth, and the width of the canal at the level of water 22 metres. In regard to the acquired increase in the width of the channel in the river and lake to 100 metres, the only object in view seems to have been to obtain an increase in the estimated cost of the work by about $10,000,000. With the very efficient and timely assistance of Lieutenant-Commander Gorringe, United States Navy, I was able to complete the computations required on the night of that day, which I delivered to the sub-committee in a tabular form on the morning of the following day. In view of the increase in the estimates made necessary by the proposed change in the width of the channel in the river and lake, I wrote a note to the first sub-committe, stating that by raising the dam proposed at Castillo Rapids one metre (and there was nothing that would make this change objectionable) the estimated cost could be reduced by no less than $14,000,000. Messrs. Cotard and Lavalley, members of the first sub-committee and engineers of great reputation, thought very favorably of the suggested change, and so reported it to the Technical Committee.

Lieutenants Wyse and Reclus had been before the sub-committee, advocating such modifications to their scheme as they thought might be accepted as a solution of the problem. Three different changes, equally objectionable, were presented during the last two days. It was at last decided by the sub-committee on locks that the canal should be provided with a tide-lock on the Pacific side and that *new channels* be made for the River Chagres and its tributaries from Matachin to the sea. With these modifications the canal without locks was reported favorably upon, and estimates of cost prepared by the first sub-committee. No surveys had been made to determine the possibility of locating the proposed new channel, and I venture to say that if such a work is ever undertaken it will be found to be of more difficult execution than is anticipated by the authors of the project.

To divert a stream from its natural course and canalize it in an artificial channel is always a work requiring mature consideration by the engineer, but when the stream desired to be controlled is a torrential river of the dimensions and conditions of the Chagres, running in close proximity to the proposed canal, which would be about forty feet below the bed of the river, then such work may be regarded as utterly impracticable.

On the 26th of May the sub-committee submitted their Reports on the different routes under consideration. A system of three locks was recommended and estimated upon by the Committee on Locks, for the Nicaragua Canal. The number of locks was fixed at 17, and the cost of each at 7,000,000 francs.

Both sub-committees reported very favorably on the Nicaragua Canal, stating that the project had been studied with much care and skill, and that the sub-committees were of the opinion that it could be executed without material difficulties.

Of the canal *à niveau* through Panama, they said that the work proposed presented such apparent difficulties of construction and so many doubtful elements, that they had been unable to agree as to its probable cost or time required to do the work.

The Nicaragua Canal, with the modifications introduced, was estimated as follows:—

	Francs.
Estimated cost of the work,	471,253,183
Twenty-five per cent. for contingencies,	117,808,290
Expenses of commissions, etc.,	29,458,329
Interest for six years, during construction,	93,000,000
	711,519,802

The canal by Panama was estimated to cost about 1,044,000,000 francs, without allowing any compensation to the Panama Railroad, which may be estimated at 200,000,000 francs, and the time of construction was fixed at twelve years.

Other projects were also estimated and reported upon; but as they were regarded as possessing but little merit, I have thought proper not to mention them here.

A general and animated discussion followed this Report, the friends of the Panama Canal claiming that the estimated cost of the Nicaragua line had been fixed too small as compared to that of Panama.

Mr. Fourcy recommended, and the President decided, that those who had presented projects should not be allowed to take part in the discussion.

It was frequently stated by Mr. Fourcy and his friends that the question of cost should have no weight in the decision of the Congress which had been called to decide as to technical possibilities.

On motion of Mr. Dauzats, supported by Mr. Fourcy, it was agreed by acclamation that the time of construction of the Nicaragua line should be raised from six years (as recommended by the subcommittees) to eight years. The estimates were also in the same informal manner raised, that of Nicaragua to 900,000,000 francs, and that of Panama to 1,200,000,000 francs.

A new plan for a canal by Panama was on that day submitted by Lieutenant Wyse, through Mr. Lepinay. It proposes the construction of an immense dam across the valley of the Chagres River, which would by this means be converted into a large artificial lake, the recipient of the waters of the Chagres and its tributaries to a point on the river called Boliro Soldado, where the dam is intended to be located. This lake was to be taken as the summit-level of the canal, and twelve locks are proposed to overcome its elevation above the sea.

This plan is an imitation of Mr. Blanchet's project for a canal by Nicaragua, and is based on the same kind of information as to natural conditions; namely, absence of surveys for the location of the dam or of the locks proposed; the extent of country to be inundated is entirely unknown, and consequently no calculations have been made to ascertain whether or not the waters of the Chagres and its tributaries coming into the lake would be sufficient in the dry season to supply the lockage and evaporation and leakage in the basin and canal. From my knowledge of those rivers, I would feel safe in stating that during two or three months in the year the supply would be found altogether insufficient, while in the rainy season much apprehension may be felt for the proper disposition of the surplus.

The sub-committees were instructed to submit estimates for that new scheme and to report the next day.

On the morning of the 27th instant Commander Selfridge called the attention of the Committee to the fact that about twenty years ago a commission of engineers had been appointed by the Government of the United States to examine and report on the then alarming condition of the harbor of Greytown, and that said commission had reported to the effect that the restoration of the harbor to its previous condition was a problem involved in much doubt; that it could not be accomplished without a very large expenditure of money, and that even then its permanence could not be assured. He also stated that Nicaragua was subject to frequent and severe earthquakes, and the territory between the lake and the Pacific was the centre of the volcanic region of Central America. He wanted to call the attention of the Committee to the injury that might result from such causes to a canal with locks in that country, and concluded by stating that Panama and the vicinity of Napipi were free from such dangers.

To this I thought proper to reply, that the commission of engineers he had referred to had based their conclusions and recommendations on the supposition that the sand-bank obstructing the entrance to the harbor of Greytown was the bar of the River San Juan; this I had already au occasion to show to the Committee was not a fact. As to the earthquakes and volcanic condition of Nicaragua, I remarked that earthquakes in Nicaragua were certainly of frequent occurrence, but that I was not aware that they had ever done any damage to life or property; that, on the other hand, it was a known fact that the old city of Panama had been totally destroyed by a shock of earthquake, and not more than two years ago a whole town had been leveled to the ground in the province of Santander, in Colombia, in which more than 15,000 lives were lost. In regard to the damage that might be done by such convulsions to a canal with locks as that proposed *via* Nicaragua, I thought they would be of less importance than what should be expected in a canal with tunnel or cuttings over three hundred feet in depth, as were proposed for the Atrato, Napipi, and Panama routes.

A general discussion as to the advantages and the disadvantages of the different projects continued all day,—Messrs. Fourcy, Dauzats, and others in favor of Panama, and Messrs. Cotard, Lavalley, and Ruelle against it.

On the morning of the 28th the sub-committee reported estimates for the new scheme of inundation proposed by Panama, amounting to 700,000,000 francs.

I then requested to be informed by the Committee whether or not the design of such a canal had been based on any actual survey or examination of the ground to determine its practicability, and that, if no surveys had been made for that purpose (as I had reason to believe), on what data had the sub-committee based the estimates; and what importance could I attach to its figures. I thought proper to add in that connection that we had been directed to present before this Congress all the information relating to the inter-oceanic canal question in possession of the Government of the United States, obtained after many years of well-conducted surveys and a large expenditure of money; that we expected to find here information of the same character; and that, from a proper comparison and discussion of all the reliable data thus obtained, competent engineers would be able to decide intelligently as to the best route for a canal. Instead of that, the only reliable and well-digested plans presented had been those from the United States, and I was sorry to see that they were weighed on the same scale with imaginary projects traced on imperfect maps of the isthmus, some of them the result of one night's inspiration.

Some confusion was produced by these remarks, and Mr. Fourcy replied that a canal with locks by Panama was no new idea, as Mr. Garalla had proposed one by that system as early as 1843. In fact, no reply was made to my inquiries.

The discussion lasted until late in the evening, Mr. Fourcy speaking for several hours in favor of a canal without locks, no matter at what cost.

It was at last agreed, amid great confusion and excitement, by a vote of 16 yeas, 11 abstentions, 3 nays, and 7 absentees, that "The Committee, standing on a technical point of view, is of the opinion that the canal, such as would satisfy the requirements of commerce, is possible across the Isthmus of Panama, and recommend especially a canal at the level of the sea."

A similar resolution was on the following day adopted by a general meeting of the Congress.

The vote was—yeas, 72; nays, 8; absentees, 37; and abstentions, 16. Of the affirmative vote only 19 were engineers, and of this last number eight are at present, or have been, connected with the Suez Canal; five are not practical engineers, and only one has been in Central America. He is a young graduate, of little or no professional experience, a native of Panama, and has assisted Lieutenant Wyse in his explorations. Of the five delegates of the French Society of Engineers, two voted *no*, and three absented themselves from the two last sessions of the Committee and the Congress.

I abstained from voting for the reason that the resolution is indefinite as to what system of canal should be finally adopted. I believe that a canal at the level of the sea, as proposed, is impracticable, at least in a commercial sense, but a canal with locks by Panama has been shown to be feasible by the United States Surveying Expedition of 1875, although more expensive than that *via* Nicaragua.

By these proceedings the remarkable condition is presented of engineers designing and estimating on the cost of such important work as the one under consideration without a proper knowledge of the ground on which the works are to be constructed, which was generally regarded, as well as that of cost, to be a matter for after consideration.

One point, however, has been gained by the discussion, viz: That the surveys made by the United States Government are now well known in Europe, and the relative merit of the different routes well appreciated by many engineers of experience and great reputation, as is shown by the Report of the first sub-committee, a copy of which I respectfully submit herewith.

It is expected that the impracticable scheme proposed by Panama will soon be abandoned for want of supporters.

I am, sir, very respectfully, your obedient servant,

A. G. MENOCAL,
Civil Engineer, United States Navy.

Hon. WM. M. EVARTS,
Secretary of State, &c., &c., &c.

APPENDIX.

PROCEEDINGS IN THE GENERAL SESSION OF THE CONGRESS IN PARIS, MAY 23, AND IN THE FOURTH COMMISSION, MAY 26, 1879.

A MEMBER.—I see that the Congress is drawing to its close. It will soon be called upon to decide between the different projects which have been submitted to it.

Shall the canal pass through Lake Nicaragua, or by way of Panama?

I beg leave to recommend the Atrato project to the attention of the President; leaving out the matter of expense and its technical question, I believe that it would have advantages for the interests of the country.

THE PRESIDENT.—I would ask that you will be kind enough to present your observations to the Commission charged with the examination of the question.

The representative of the Fifth Commission has now the floor.

THE SECRETARY OF THE FIFTH COMMISSION.—The Fifth Commission had announced its Report for to-day's session; it met yesterday, and, after discussion, became convinced that it would be impossible to hand in its Report before being fully advised of the decision which the Statistical Commission will adopt. This Commission has informed us that they will complete their decisions at the sitting of to-morrow (Saturday). We have been obliged, therefore, to adjourn until the communication of the decisions which will be made at that session.

THE PRESIDENT.—We will have a general session, then, next Tuesday, at nine o'clock, to hear the different Reports before the closing session, which will certainly take place on Thursday evening at nine.

Now, gentlemen, permit me, as the President, to make a *résumé* of what has been done at the session of the Congress. You have heard the different Reports which have been laid before you. What strikes us the most is the enthusiasm of the United States of America in favor of locating a canal at Panama; they have been at considerable expense in sending out explorers to that isthmus. It is unquestionable that America has hereby given us proof of her impartiality, and of her devotion (to the interests of a canal). Let no one, hereafter, come to tell us that she does not wish a canal because she has a railroad. Rear-Admiral Ammen and Commander Selfridge, in their study of the projects, have expressed to us the desire to come to an understanding in order to secure something practicable.

Commander Selfridge has presented to us his plan for the Napipi-Atrato route,—a plan without locks, except the two by which to descend from the level of the Atrato to the level of the Pacific. M. Wyse and his companions have reported on the mission which they undertook. Of the seven who went out, four died in those wilds where one cannot advance a step except with hatchet in hand. They have at length returned, and have had the frankness to declare to us that in the countries which they have just explored a canal is impossible. When I saw them, it was in company with Monsieur Lavalley, our illustrious Suez engineer, who has already designed for us so many machines, and who, in like circumstances, will well know how to invent new ones. I have consulted M. Lavalley, and he has replied that the decision would be for a canal à *niveau*—that this would be the general opinion. I shall permit myself to sustain this opinion. Although I am not an engineer, my experience has often served me in this matter. It is very necessary to decide quickly; if we should occasion delay, we will be the cause of great injury to commerce.

The most distinguished engineers of France have written some articles on this subject, which you have been able to read in the *Revue des Deux Mondes*. Some of them are in favor of a canal with fifteen locks; that is to say, twice as many as any one has already constructed in a work of this kind, and this for the simple reason that they think it impossible to construct a tunnel. Moreover, I learned a fortnight ago, that as the result of the soundings very recently made we cannot count upon an outlet from the canal into the Pacific; but, at this day, when dredgers have come into use, we have learned by the regime of waters that what could not be done some years ago is now perfectly practicable. It would have happened so to us at Suez (*i.e.*, we could not have had an outlet there) if we had left our harbors without cleaning them out. God only knows at what we should have arrived in a century or two more; we have learned now to secure the equilibrium of the waters.

I do not wish to interpose the least difficulty in the labors of the Technical Commission. The savants have the means of foreseeing very many things, but there are some facts which are incontrovertible.

APPENDIX. 97

I will ask the Commission not to formulate any resolutions which could arrest certain plans. I would wish them to say only yes or no, whether it appears to them possible to construct a canal à niveau, which seems to be the desideratum of the whole world. M. Lavalley has studied that question of a tunnel; he believes it clearly practicable. It is, he says, only a question of expense. I will not enter into the scientific question. I will only ask the Technical Commission to tell us precisely what would be the expense of a canal à niveau; what estimates can be made of that expense, and especially what the cost will be in the future (after construction) for canals à niveau, or for those with locks. Governments can encourage such enterprises; they cannot execute them. It is the public, then, on whom we must call, and when you come before them they will ask of you (if it is a canal with locks), What will be the expense in the future?

I will express my opinion. I consider that a canal with locks would retard navigation. From the experience we have had at Suez, a ship must not now be delayed. There are a thousand Philistines there who can load a ship of four hundred tons in an hour. It is to be remembered, also, that it costs a ship of a thousand tons two thousand francs for every day's delay. I have often been consulted on this subject; I have always replied, by dispatch, that ships must not be delayed in their passage; they are informed that it will be well for them to wait the rise of the tide when they cannot be sure of their being able to steer satisfactorily. The larger number of them, in spite of the delay, prefer to wait several hours for the tide. This is certainly an obstacle, and we thought of remedying it at Suez by means of tide-locks; but we have been obliged to abandon that idea. I go every year to Suez. I have often met there with large ships which were passing with the ebb tide because they were in haste; I have witnessed the slowness of their sailing. We must not forget that the large vessels which come from England must make their voyage in twenty-one days. I have often seen these English vessels of twenty-five hundred tons sheltering an entire regiment. These brave people are on the deck with their wives and children, for you know they never travel without them. We look on them from a distance; they look like anything but soldiers.

In my opinion we ought not to make a canal at Panama with locks, but à niveau; that is the opinion of the public, whose organ I am.

Don Manuel Peralta.—I have listened with the most profound respect to the opinion which Monsieur de Lesseps has just expressed. It appears to me he has forgotten to say that the Government of the United States appointed a Scientific Commission of three members, and that these gentlemen successively examined the different projects from the route by the Atrato to the passage through the great Lake Nicaragua, as well as the canal à *niveau*, with a tunnel alongside the railroad from Colon to Panama; but after a long and very conscientious study, and repeated explorations, they decided that the plan of Commander Lull was preferable from a humanitarian point of view, since it would save a great many lives, a circumstance which had not been foreseen when the Panama Railroad was built; and, besides that, by this route sailing-vessels could always pass with favorable wind, which cannot be hoped for on other routes, especially at Panama. The American Scientific Commission decided on the Nicaragua route, since it was demonstrated that the Panama route was impossible. Is there no ground for taking its decision into consideration?

The President.—The Government declared that there was no impossibility for a canal with locks, and that it *could not* assert that a canal à *niveau* was impossible. The Government, in its fairness, was absorbed in one idea only. It acted on the principle that nothing was impossible.

M. Des Grand.—Since a Commission has been named composed of members of the bureaus of the different sections, would it not be well that this Commission meet to formulate a programme of questions to be resolved,—a programme which can be distributed in time for us to reflect upon them before the last sitting of the Congress. In this way, instead of being limited to the one question, Which is best of all the canals?—or which is the same as asking Which is the best of all the Governments?—we can enlighten the public more by increasing the number of questions. I wish to say, that what is understood to be the best of all the canals is not, perhaps, that which costs the least.

The President.—The different sections can send in their memoranda to the Technical Commission, which will examine all their plans.

M. Des Grand.—If we have before us this question only—Which is the best of the canals?—I fear we shall not reach its solution.

The President.—We shall reach it in time.

APPENDIX. 99

M. DES GRAND.—I address the Central Commission, composed of the Presidents and Secretaries of all the Committees, and I ask it to formulate a programme to which the Technical Commission will have to reply.

THE PRESIDENT.—We cannot impose a programme on that Commision.

M. RUELLE.—I will limit myself to those questions which I address to the Commission of Navigation to facilitate the labor of the Technical Commission. We have been told up to the present time that these would be tide-locks. I wish the Commission of Navigation to pronounce clearly on this point, since on the construction or the non-construction of tide-locks may depend a difference of height in the tunnel. Then a second question as to the form, whether ovoid or elliptical, which of these is most economical for constructing a tunnel? and in which does the arch offer the most resistance? I wish to know if the width necessary for the passage of sailing-ships can accord with the elliptical form, and if we must be limited to a width of 22 metres for the tunnel.

My third question to the Commission of Navigation is this: What comparison can be made between the difficulties of a ship passing through several locks and the passage of the same ship through a tunnel of a certain length. In a word, will a tunnel ten kilometres long present the same obstacles to navigation as the sailing of a ship through nine or ten locks.

M. VOISIN NAY.—From the questions which M. Ruelle addresses to the Commission of Navigation, I wish to set aside the first one; that is, whether guard-locks ought to be constructed on the Pacific side. I add that this question can be answered only among ourselves.

M. SIMONIN.—As the Congress must close its sessions next Thursday, and as our honorable President has just told us that we must examine the cost of a canal à *niveau*, whatever it may cost, I ask permission to present some observations precisely on what our President has just said. We, who have met in this body, are not only engineers, geographers, theoretical scientists; it has seemed to me that we are also manufacturers, seamen, and business men, and in the preceding general session our honorable President said that in this matter it was not necessary to have the assistance of Government, and that we must address ourselves to the public only.

Now it has been announced that the transit of the future canal will reach the figures of six millions of tons, which I consider a very large maximum; and I have the right to so regard it, if I compare it with the figure of three millions of tons passing through the Suez Canal, which is an important point of the world's commerce. Whatever may be the importance of the Panama Canal, I doubt whether it will ever exceed that which the Suez Canal has attained. I doubt it for more than one reason, because the Suez Canal leads to China, which represents more than seven hundred millions of people, while in America you can only count on markets representing scarcely one hundred million. Including countries as far as the Straits of Magellan, you will not have the millions of China. I believe, then, that the figure of six millions of tons which you have fixed upon as an average is exaggerated. Since we ought to treat it simply as a business affair, we should correct this estimate. We are going to make it an affair of the long future, as M. de Lesseps said, but I believe that this would be to proceed on a very bad principle; we are going to adapt the canal à niveau, cost what it may, even though it should be five millions, the estimate which has been made by a person of great authority in the United States. I say that we should examine the question from another point of view; we should inquire what a canal with locks might be, besides asking specialists how many ships could pass through these locks in 24 hours: whether a sufficient number could pass for the number of tons which has been just spoken of. I am not the advocate of any company or any canal; I have no reason to adopt one system rather than another, but I do not wish that we should go out from this body with a negative vote. If we decide for a canal which would be impracticable, it would be unfortunate. We ought to treat the question as business men. It is necessary that this canal shall give large dividends to its stockholders; without that we shall have played a singular part. I desire that our honorable President will tell me if I have misunderstood him when he asked the Technical Commission to decide for a canal at any cost provided it was à niveau.

M. DE LESSEPS.—I have said that the Commission would pronounce for one canal or another. We must decide. We ought not to go out from this body without making a decision.

M. SPEINENT.—I would remark to the assembly that we were already acquainted with the opinion of M. de Lesseps. He said, in the Geographical Congress of 1875, that he had always advocated the use of canals à niveau everywhere where their construction had been shown to him possible. I believe he thinks a canal à niveau would be preferable via Panama; if that should be impossible there, we must go to Nicaragua.

APPENDIX. 101

M. DE LESSEPS.—Why not a fresh-water canal.

M. FONTANE.—I ask leave to make a single observation in reply to M. Simonin. He said that he accepted the figure of 6,000,000 tons having to pass through the inter-oceanic canal as a maximum. I wish simply to appeal to the patience of M. Simonin, and ask him to await the Report which the President of the Statistical Commission, M. Levapeur, will present in a few days to the assembly. He will find in this Report all the elements necessary to form a definite opinion. He added that he accepted the figures of 6,000,000 of tons as a very large maximum, because there were but 3,000,000 of tons which passed through the Suez Canal. I desire to recall to the Congress that, at its first meeting, in a Report which I had the honor to present, I stated precisely the reason why the Suez Canal has now a traffic of 3,000,000 tons. It is simply because steam navigation almost alone can properly utilize the Suez Canal, and the shipbuilders cannot make enough steam vessels to carry on the traffic which exists. I showed by official figures that the traffic between the Western and Eastern world through the Suez Canal would reach 10,000,000 tons; consequently the figure of 6,000,000 tons that M. Simonin considers as a maximum for an inter-oceanic canal, comparing it with the 3,000,000 tons of the Suez Canal, becomes almost a minimum, compared with 10,000,000 which could pass through the Suez Canal.

A second opinion of M. Simonin bore upon the sum that the Panama Canal might cost, a sum which is reckoned at some milliards! We would like upon this point again to appeal to the patience of our honorable colleague, and ask him to await the developments of the Report of the Technical Commission, charged with estimating the cost of the canal. I have wished simply to appeal to the patience of M. Simonin. It is perhaps asking too much.

M. SIMONIN.—I do not see why, on two occasions, M. Fontane, my honorable colleague and friend, has made an appeal to my patience. It seems to me that I speak with much calmness, and that I have always given proof of much patience and tranquillity; consequently, M. Fontane has wished to be witty, but I do not admit his objection. As to the figures that I have made use of, I would be glad that he shall be good enough to demonstrate to me that they are exaggerated. We are doing a work in the interest of humanity. I will experience great satisfaction when 10,000,000 tons will have to pass through the canal we desire to open.

The figures that I have cited have been made by conscientious persons; show me that we are wrong—I ask nothing better; but it was useless to occupy the attention of so numerous and busy an assemblage in making personal question.

M. DE LESSEPS.—There is nothing personal in that question (in what has been said); it has only been the occasion for us to learn the opinions of those in listening to whom we have always great pleasure. M. Simonin has said that we have ten francs since the Suez Canal has been opened to navigation, and that considering the number of vessels which pass through the canal we had a return of ten per cent. In truth, we have already received large returns, and we have been obliged to reduce our tariffs.

Through whatever route the Panama Canal will pass, it will shorten the distance more readily than the route of the Suez does. During the forty years that I have studied the question I have always understood that for a profit it is necessary to receive at the least ten francs per ton; we can readily make the American Canal pay double of this, whatever may be the project that is brought about. These are considerations that one is very glad to know for the future. Commander Selfridge makes known to the assembly, through the interpreter, that the people of the United States have no preference for any project. Every one in America is convinced of the special competence of the Congress, and whatever may be the result, the nation will accept with eagerness all the decisions which will be taken here. (Applause.)

M. LAROUESSE—I would ask permission of the Congress to put another question, in addition to those of M. Ruelle; it is to compare the difference of distances which the proposed canal would give, considering the principal regions which they would be called upon to benefit. It is certain that the Congress, in order to make a decision, shall have also to estimate the time judged necessary to pass through the different canals, etc., all of which I think require some explanation.

M. EIFFEL.—I desire that the Commission on Navigation will be good enough to give us some explanation upon the inconvenience of locks, if the number of them were reduced. In other words, would canals with locks present much less inconvenience with less than half the number of locks than by the [canal plan] as now presented?

M. DE LESSEPS.—If no other person wishes to speak, I will remind the assembly that our first general meeting will take place Tuesday next at 9 A. M., to hear the reports of the First and Fifth Commissions, and that the closing session is fixed for Thursday.

The adjournment took place at 6.45 P. M.

www.ingramcontent.com/pod-product-compliance
Lightning Source LLC
Chambersburg PA
CBHW020859160426
43192CB00007B/997